Third Grade Math Puzzle Worksheets

by: Laura Putman, Bright Minds Engaged

© Laura Putman, Bright Minds Engaged, 2023-present, All rights reserved.

© Laura Putman , Bright Minds Engaged, 2023-present, All rights reserved.

All images created by Laura Putman Digitals LLC. All rights reserved. No part of this publication may be reproduced, distributed, or transmitted in any form or by any means. This includes photocopying, recording, or other electronic or mechanical methods without prior permission of the publisher, except in the case of brief quotations embodied in critical reviews and other
noncommercial uses permitted by copyright law.

Clipart by Educlips – www.educlips.com

No part of this product maybe used or reproduced for commercial use.

Contact the author :
laura@thirdgradeengaged.com

PUZZLE 1

Directions: Cut the puzzle pieces from the next page apart.
To solve the puzzle, glue the piece with the matching answer on each problem.

What is the value of the underlined digit? **6**,294	Which digit is in the hundreds place? 7,049	What number is missing? 2,871 = 2,000 + ____ + 70 + 1
What is this number in standard form? three thousand, two hundred	What is the value of the underlined digit? 5,1**6**7	Which digit is in the ones place? 1,492
What number is missing? 4,487 = 4,000 + 400 + ____ + 7	Which digit is in the thousands place? 6,240	What is this number in standard form? 5,000 + 500 + 10 + 1
How many tens are in this number? 8,652	What is the value of the underlined digit? 6,8**3**4	What is missing? 3,754 = three thousand, seven hundred _____
What is this number in standard form? one thousand, thirteen	Which digit is in the tens place? 3,246	What is this number in standard form? 5,000 + 100 + 10 + 5

©Laura Putman, Bright Minds Engaged, 2023-present All rights reserved.

©Laura Putman, Bright Minds Engaged, 2023-present All rights reserved.

PUZZLE 1

Directions: Cut these puzzle pieces apart. Glue the piece with the answer on top of each matching problem on the previous page.

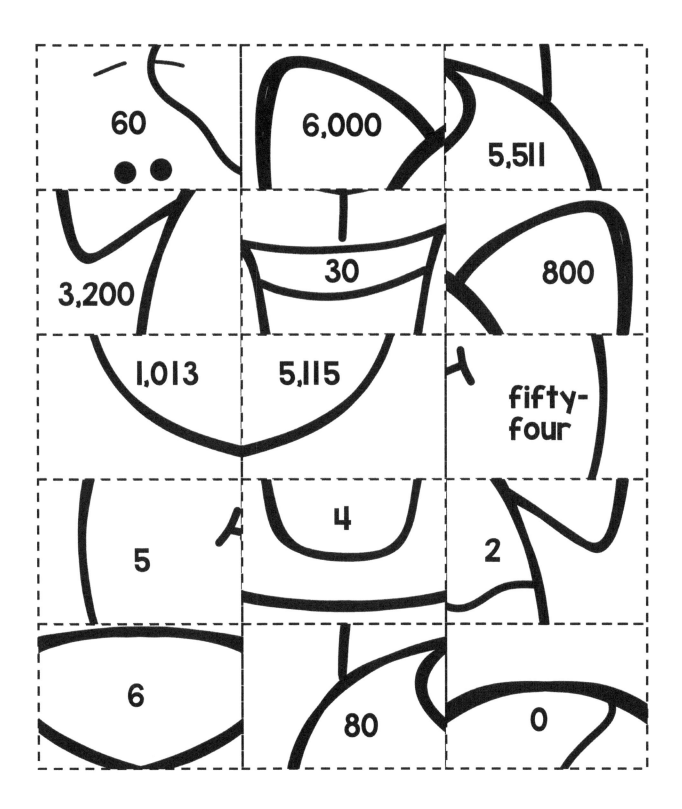

©Laura Putman, Bright Minds Engaged, 2023-present All rights reserved.

©Laura Putman, Bright Minds Engaged, 2023-present All rights reserved.

PUZZLE 2

Directions: Cut the puzzle pieces from the next page apart.
To solve the puzzle, glue the piece with the matching answer on each problem.

What is this number in standard form? 1,000 + 500 + 2	How many thousands are in this number? 1,048	Which digit is in the tens place? 3,542	What is missing? 5,702 = five thousand, seven ____ two
What is the value of the underlined digit? 8,7<u>1</u>3	Which digit is in the thousands place? 5,792	What number is missing? 9,763 = ____ + 700 + 60 + 3	What is this number in standard form? one thousand, fifty-two
What is this number in standard form? 5,000 + 100 + 20 + 2	How many hundreds are in this number? 6,347	Which digit is in the ones place? 2,768	What is the value of the underlined digit? 7,<u>4</u>02
What is missing? 4,565 = four ____, five hundred sixty-five	Which digit is in the hundreds place? 4,607	What number is missing? 3,285 = ____ + 200 + 80 + 5	What is this number in standard form? two thousand, five

©Laura Putman, Bright Minds Engaged, 2023-present All rights reserved.

©Laura Putman, Bright Minds Engaged, 2023-present All rights reserved.

PUZZLE 2

Directions: Cut these puzzle pieces apart. Glue the piece with the answer on top of each matching problem on the previous page.

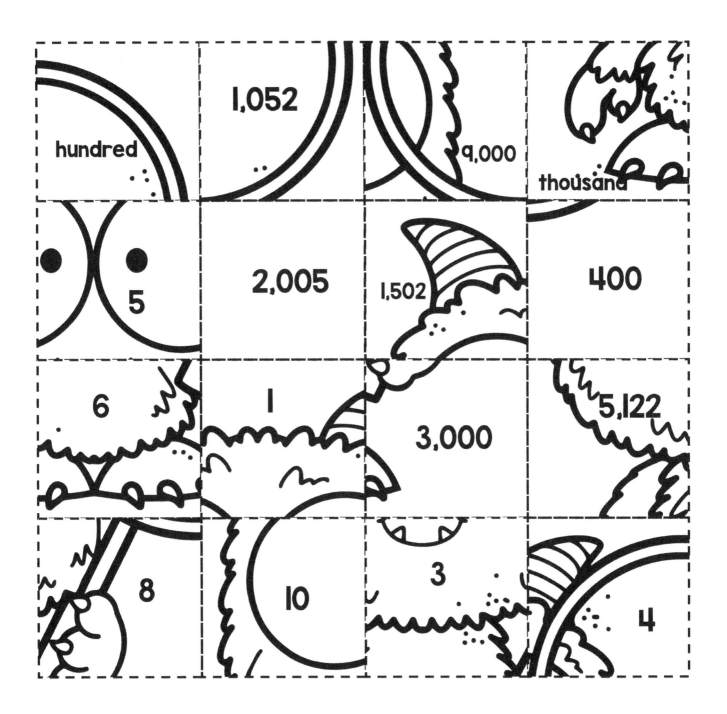

©Laura Putman, Bright Minds Engaged, 2023-present All rights reserved.

©Laura Putman, Bright Minds Engaged, 2023-present All rights reserved.

PUZZLE 3

Directions: Cut the puzzle pieces from the next page apart.
To solve the puzzle, glue the piece with the matching answer on each problem.

Round to the nearest ten: 476	Round to the nearest hundred: 5,021	Round to the nearest ten: 89	Round to the nearest hundred: 7,835
Round to the nearest ten: 616	Round to the nearest hundred: 62	Round to the nearest ten: 3,879	Round to the nearest hundred: 3,963
Round to the nearest ten: 526	Round to the nearest hundred: 4,521	Round to the nearest ten: 308	Round to the nearest hundred: 1,529
Round to the nearest ten: 5,236	Round to the nearest hundred: 6,846	Round to the nearest ten: 18	Round to the nearest hundred: 8,417

©Laura Putman, Bright Minds Engaged, 2023-present All rights reserved.

©Laura Putman, Bright Minds Engaged, 2023-present All rights reserved.

PUZZLE 3

Directions: Cut these puzzle pieces apart. Glue the piece with the answer on top of each matching problem on the previous page.

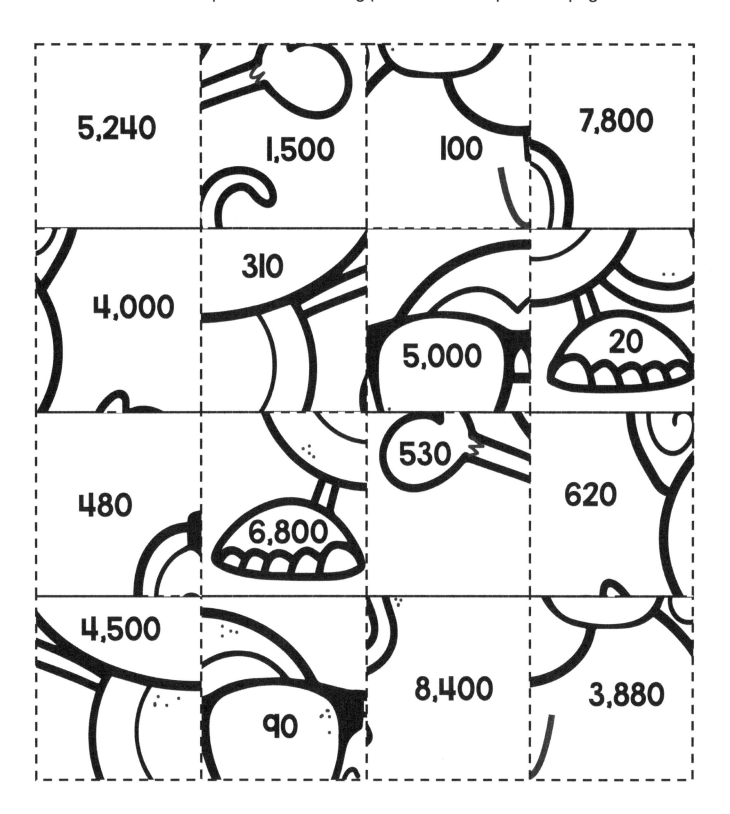

©Laura Putman, Bright Minds Engaged, 2023-present All rights reserved.

PUZZLE 4

Directions: Cut the puzzle pieces from the next page apart.
To solve the puzzle, glue the piece with the matching answer on each problem.

Round to the nearest ten: 547	Round to the nearest hundred: 6,728	Round to the nearest ten: 1,459
Round to the nearest hundred: 645	Round to the nearest ten: 89	Round to the nearest hundred: 1,834
Round to the nearest ten: 236	Round to the nearest hundred: 9,343	Round to the nearest ten: 761
Round to the nearest hundred: 963	Round to the nearest ten: 84	Round to the nearest hundred: 5,527

©Laura Putman, Bright Minds Engaged, 2023-present All rights reserved.

©Laura Putman, Bright Minds Engaged, 2023-present All rights reserved.

Directions: Cut these puzzle pieces apart. Glue the piece with the answer on top of each matching problem on the previous page.

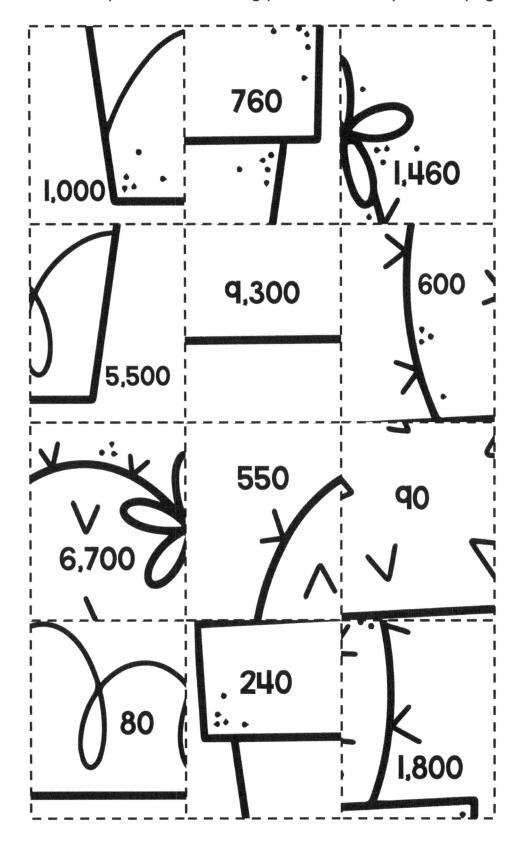

©Laura Putman, Bright Minds Engaged, 2023-present All rights reserved.

PUZZLE 5

Directions: Cut the puzzle pieces from the next page apart.
To solve the puzzle, glue the piece with the matching answer on each problem.

876 + 124	430 + 398	286 + 225
149 + 783	244 + 711	604 + 356
132 + 852	365 + 302	195 + 125
545 + 247	180 + 390	203 + 599

©Laura Putman, Bright Minds Engaged, 2023-present All rights reserved.

PUZZLE 5

Directions: Cut these puzzle pieces apart. Glue the piece with the answer on top of each matching problem on the previous page.

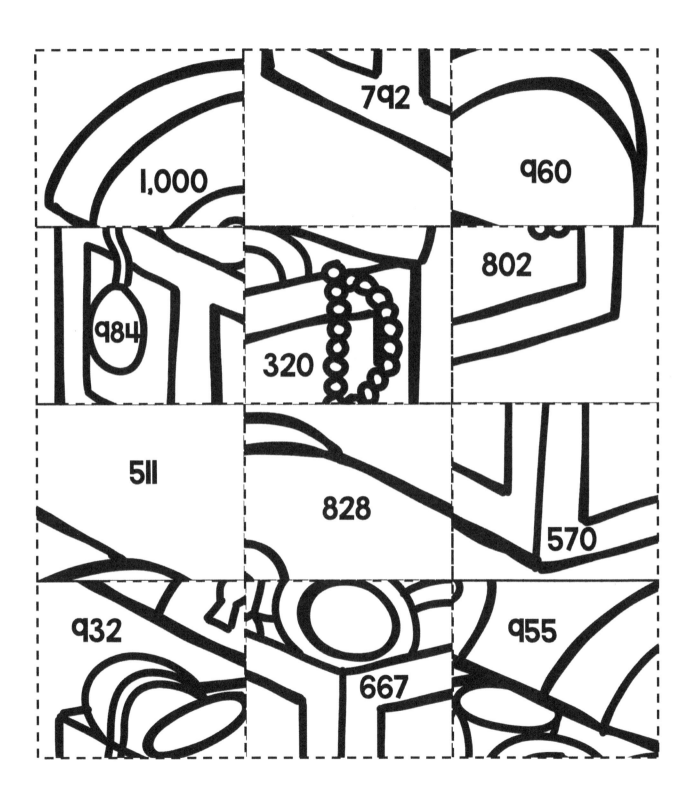

©Laura Putman, Bright Minds Engaged, 2023-present All rights reserved.

PUZZLE 6

Directions: Cut the puzzle pieces from the next page apart.
To solve the puzzle, glue the piece with the matching answer on each problem.

861 + 129	238 + 732	679 + 312
554 + 410	900 + 100	378 + 461
124 + 293	356 + 507	106 + 890
258 + 565	494 + 312	320 + 350

©Laura Putman, Bright Minds Engaged, 2023-present All rights reserved.

©Laura Putman, Bright Minds Engaged, 2023-present All rights reserved.

PUZZLE 6

Directions: Cut these puzzle pieces apart. Glue the piece with the answer on top of each matching problem on the previous page.

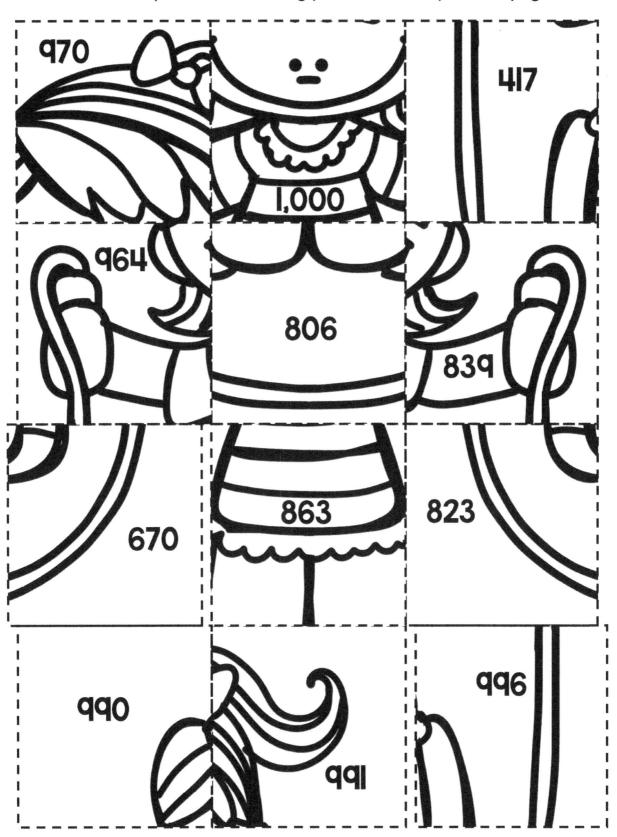

©Laura Putman, Bright Minds Engaged, 2023-present All rights reserved.

PUZZLE 7

Directions: Cut the puzzle pieces from the next page apart.
To solve the puzzle, glue the piece with the matching answer on each problem.

245 − 173	600 − 272	890 − 45	753 − 543
1,000 − 481	558 − 379	174 − 122	394 − 156
726 − 241	403 − 366	900 − 205	656 − 398
560 − 347	211 − 41	883 − 295	374 − 326

©Laura Putman, Bright Minds Engaged, 2023-present All rights reserved.

©Laura Putman, Bright Minds Engaged, 2023-present All rights reserved.

Directions: Cut these puzzle pieces apart. Glue the piece with the answer on top of each matching problem on the previous page.

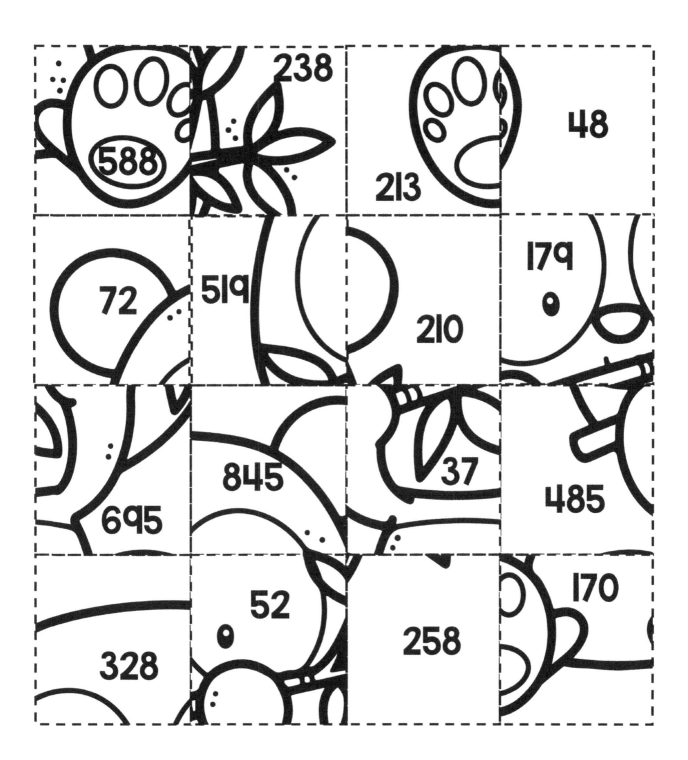

©Laura Putman, Bright Minds Engaged, 2023-present All rights reserved.

PUZZLE 8

Directions: Cut the puzzle pieces from the next page apart.
To solve the puzzle, glue the piece with the matching answer on each problem.

354 - 155	797 - 503	243 - 77
829 - 766	333 - 106	1,000 - 934
402 - 78	217 - 96	100 - 22
619 - 522	206 - 131	552 - 354

©Laura Putman, Bright Minds Engaged, 2023-present All rights reserved.

©Laura Putman, Bright Minds Engaged, 2023-present All rights reserved.

PUZZLE 8

Directions: Cut these puzzle pieces apart. Glue the piece with the answer on top of each matching problem on the previous page.

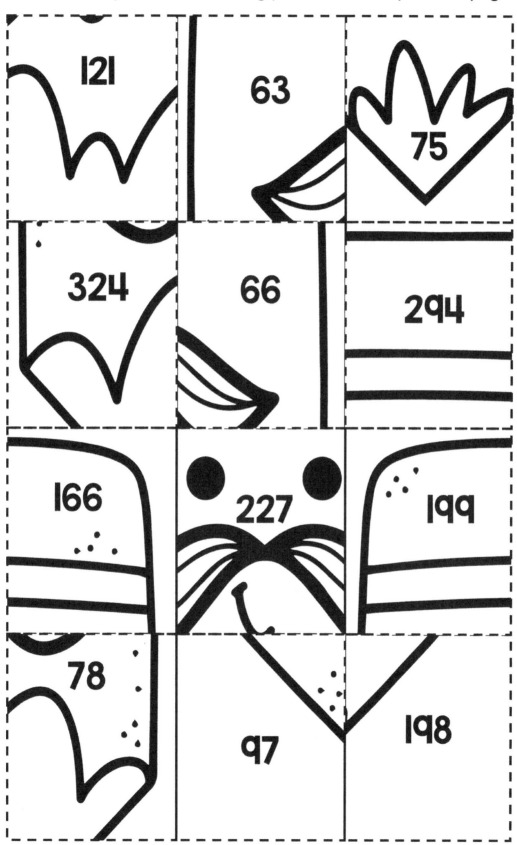

©Laura Putman, Bright Minds Engaged, 2023-present All rights reserved.

PUZZLE 9

Directions: Cut the puzzle pieces from the next page apart.
To solve the puzzle, glue the piece with the matching answer on each problem.

1,000 − 750	467 + 167	690 + 218	212 − 156
738 + 154	802 − 792	371 + 229	194 − 73
550 + 450	284 + 217	900 − 366	496 − 426
157 + 195	784 − 158	603 − 360	194 + 74

©Laura Putman, Bright Minds Engaged, 2023-present All rights reserved.

©Laura Putman, Bright Minds Engaged, 2023-present All rights reserved.

Directions: Cut these puzzle pieces apart. Glue the piece with the answer on top of each matching problem on the previous page.

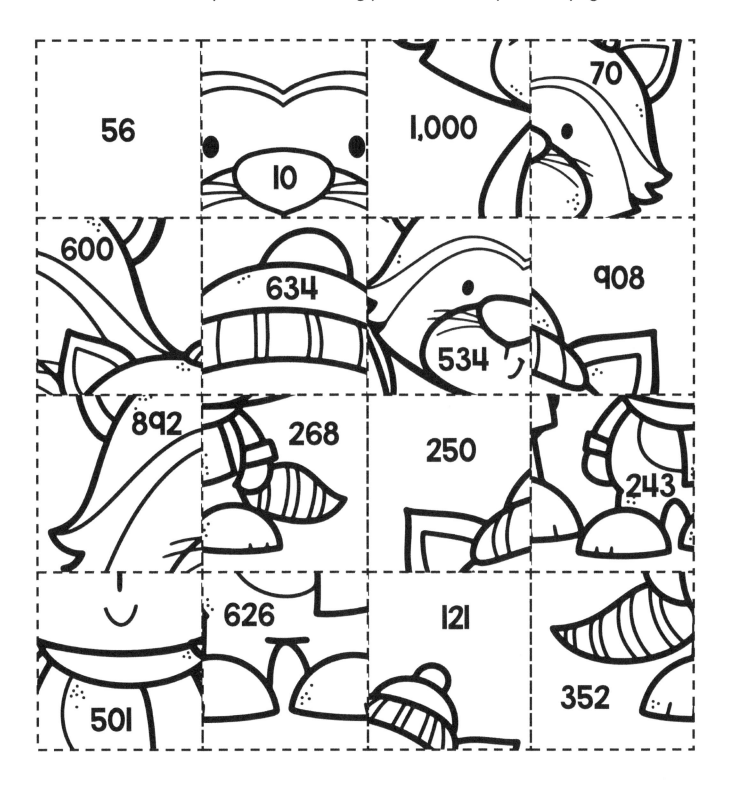

PUZZLE 10

Directions: Cut the puzzle pieces from the next page apart. To solve the puzzle, glue the piece with the matching answer on each problem.

1,000 − 435	439 + 199	825 − 117
767 + 193	148 + 35	309 − 226
216 + 18	571 − 98	900 − 622
692 + 308	453 − 163	105 + 245

©Laura Putman, Bright Minds Engaged, 2023-present All rights reserved.

©Laura Putman, Bright Minds Engaged, 2023-present All rights reserved.

Directions: Cut these puzzle pieces apart. Glue the piece with the answer on top of each matching problem on the previous page.

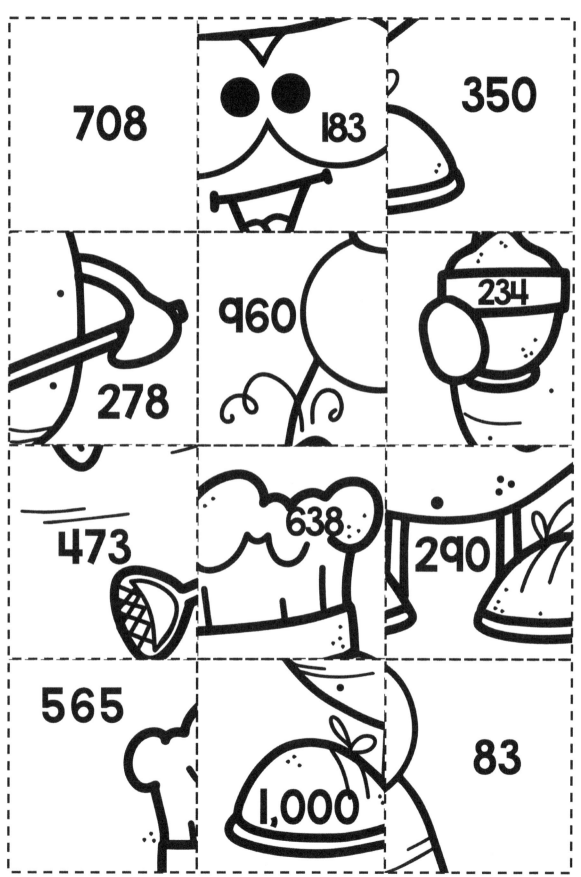

©Laura Putman, Bright Minds Engaged, 2023-present All rights reserved.

PUZZLE 11

Directions: Cut the puzzle pieces from the next page apart. To solve the puzzle, glue the piece with the matching multiplication problem on each addition problem.

8 + 8 + 8	9 + 9 + 9 + 9 + 9 + 9 + 9 + 9 + 9	1 + 1 + 1 + 1 + 1 + 1
4 + 4 + 4 + 4 + 4 + 4 + 4 + 4 + 4 + 4	7 + 7 + 7	7 + 7 + 7 + 7 + 7 + 7
8 + 8 + 8 + 8 + 8 + 8	2 + 2 + 2 + 2 + 2	5 + 5 + 5 + 5
7 + 7 + 7 + 7 + 7 + 7 + 7 + 7 + 7 + 7	10 + 10 + 10 + 10 + 10	9 + 9 + 9

©Laura Putman, Bright Minds Engaged, 2023-present All rights reserved.

©Laura Putman, Bright Minds Engaged, 2023-present All rights reserved.

PUZZLE 11

Directions: Cut these puzzle pieces apart. Glue the piece with the matching multiplication on top of each matching addition problem on the previous page.

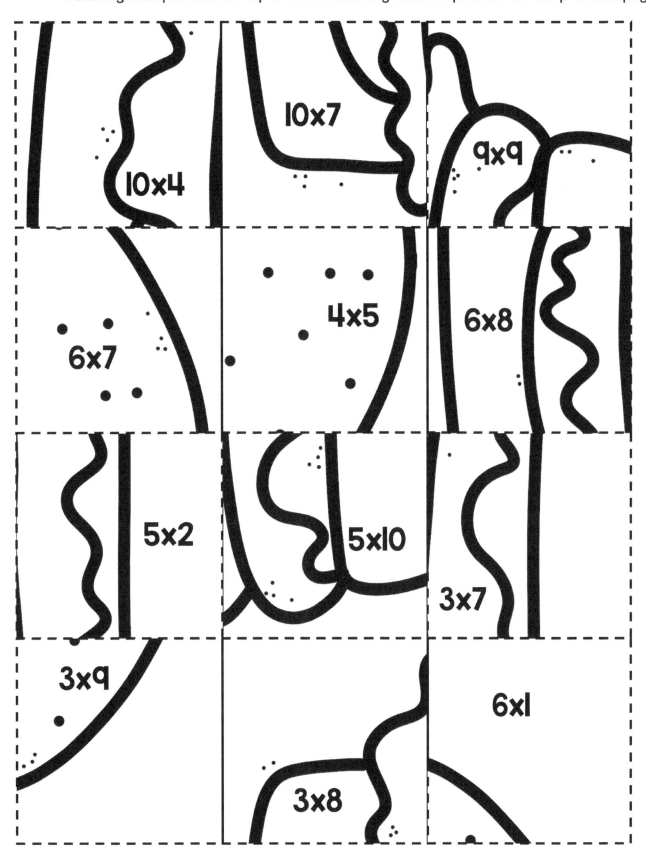

©Laura Putman, Bright Minds Engaged, 2023-present All rights reserved.

©Laura Putman, Bright Minds Engaged, 2023-present All rights reserved.

PUZZLE 12

Directions: Cut the puzzle pieces from the next page apart. To solve the puzzle, glue the piece with the matching multiplication problem on each addition problem.

1 + 1 + 1 + 1 + 1	9 + 9 + 9	7 + 7 + 7
9 + 9 + 9 + 9 + 9 + 9 + 9 + 9	7 + 7 + 7 + 7 + 7 + 7 + 7 + 7 + 7 + 7	1 + 1 + 1 + 1 + 1 + 1 + 1 + 1 + 1 + 1
6 + 6 + 6 + 6 + 6 + 6	5 + 5 + 5 + 5 + 5	3 + 3 + 3
7 + 7 + 7 + 7 + 7 + 7	8 + 8 + 8 + 8	2 + 2
1 + 1 + 1 + 1	10 + 10 + 10	5 + 5 + 5 + 5 + 5 + 5 + 5 + 5

©Laura Putman, Bright Minds Engaged, 2023-present All rights reserved.

©Laura Putman, Bright Minds Engaged, 2023-present All rights reserved.

PUZZLE 12

Directions: Cut these puzzle pieces apart. Glue the piece with the matching multiplication on top of each matching addition problem on the previous page.

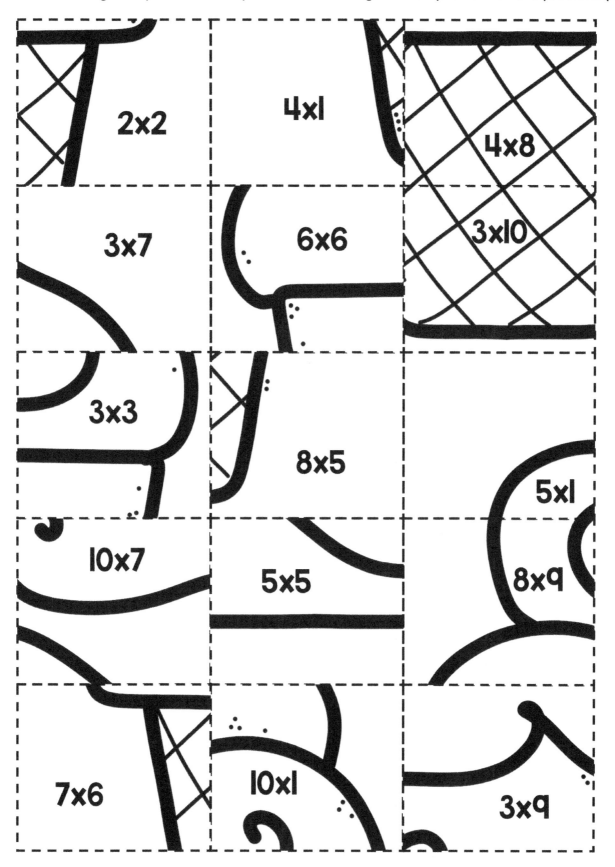

©Laura Putman, Bright Minds Engaged, 2023-present All rights reserved.

©Laura Putman, Bright Minds Engaged, 2023-present All rights reserved.

PUZZLE 13

Directions: Cut the puzzle pieces from the next page apart. To solve the puzzle, glue the piece with the matching answer on each problem.

1x2	8x5	7x4	9x6	2x2
2x9	1x1	6x5	0x3	4x8
7x10	4x4	5x9	3x1	11x10
6x7	3x11	8x8	3x2	6x8
10x12	2x4	4x9	9x11	5x7

©Laura Putman, Bright Minds Engaged, 2023-present All rights reserved.

©Laura Putman, Bright Minds Engaged, 2023-present All rights reserved.

PUZZLE 13

Directions: Cut these puzzle pieces apart. Glue the piece with the answer on top of each matching problem on the previous page.

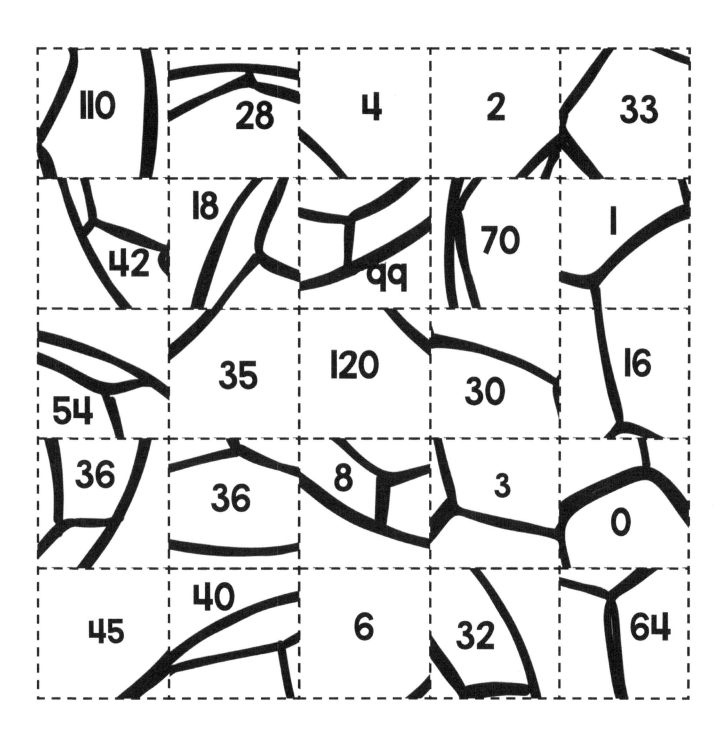

©Laura Putman, Bright Minds Engaged, 2023-present All rights reserved.

PUZZLE 14

Directions: Cut the puzzle pieces from the next page apart. To solve the puzzle, glue the piece with the matching answer on each problem.

9x6	11x1	4x8	1x7	3x2
8x7	9x2	7x10	1x5	5x5
7x4	10x3	7x5	3x12	4x6
12x0	5x4	4x4	6x2	8x9
7x7	1x2	9x11	3x3	10x8

©Laura Putman, Bright Minds Engaged, 2023-present All rights reserved.

Directions: Cut these puzzle pieces apart. Glue the piece with the answer on top of each matching problem on the previous page.

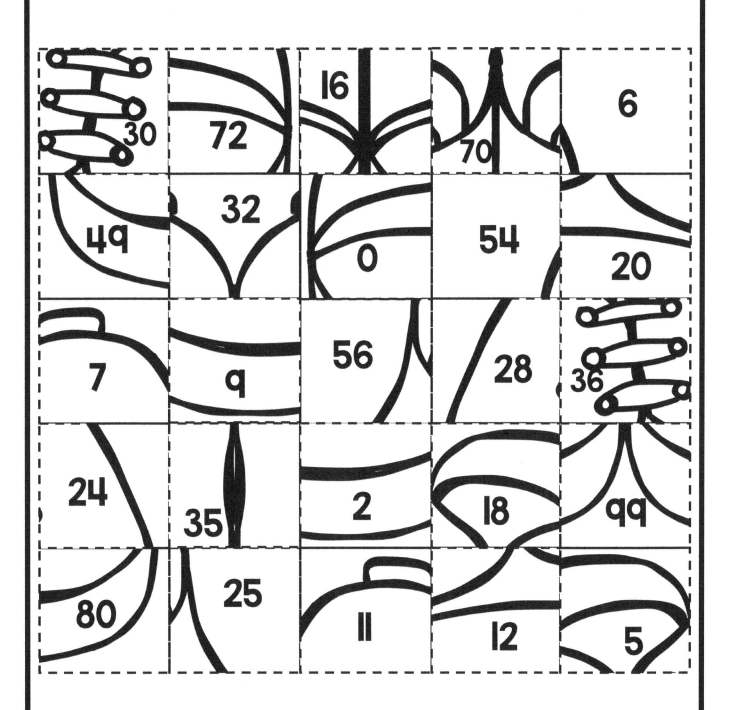

©Laura Putman, Bright Minds Engaged, 2023-present All rights reserved.

PUZZLE 15

Directions: Cut the puzzle pieces from the next page apart. To solve the puzzle, glue the piece with the matching answer on each problem.

$2 \times (4 \times 3)$	$(3 \times 2) \times 6$	$(2 \times 2) \times 4$	$11 \times (5 \times 2)$
$(3 \times 3) \times 7$	$10 \times (5 \times 2)$	$4 \times (6 \times 2)$	$(1 \times 1) \times 3$
$10 \times (2 \times 2)$	$(3 \times 3) \times 9$	$(2 \times 2) \times 8$	$10 \times (6 \times 2)$
$8 \times (4 \times 2)$	$(1 \times 5) \times 5$	$0 \times (6 \times 0)$	$(3 \times 3) \times 9$

©Laura Putman, Bright Minds Engaged, 2023-present All rights reserved.

©Laura Putman, Bright Minds Engaged, 2023-present All rights reserved.

PUZZLE 15

Directions: Cut these puzzle pieces apart. Glue the piece with the answer on top of each matching problem on the previous page.

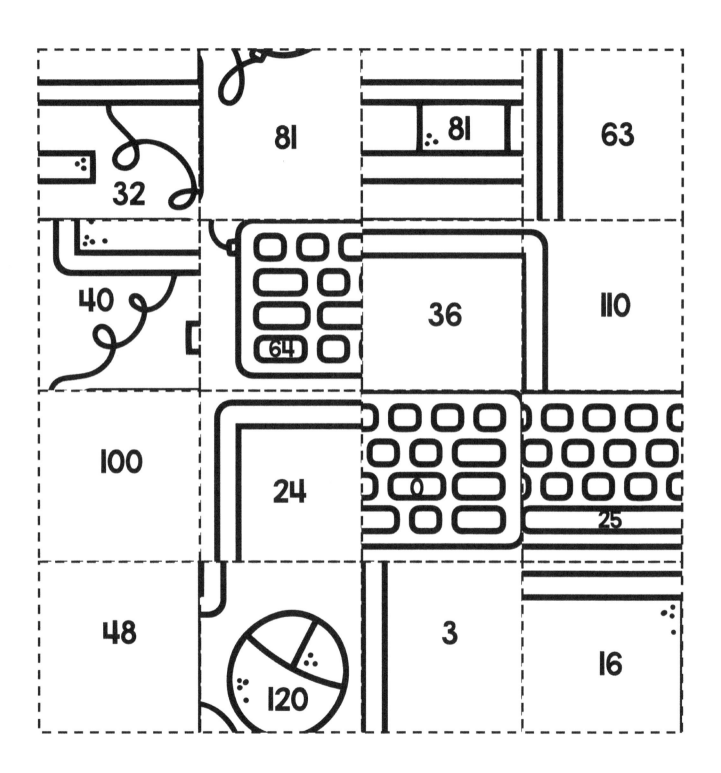

©Laura Putman, Bright Minds Engaged, 2023-present All rights reserved.

PUZZLE 16

Directions: Cut the puzzle pieces from the next page apart. To solve the puzzle, glue the piece with the matching answer on each problem.

2 x (3 x 2)	(2 x 4) x 6	5 x (5 x 2)	(2 x 2) x 6
(5 x 2) x 7	0 x (3 x 9)	1 x (5 x 9)	(2 x 2) x 7
6 x (1 x 6)	5 x (2 x 6)	(2 x 2) x 8	(4 x 2) x 8
11 x (2 x 2)	(4 x 3) x 8	(2 x 4) x 5	11 x (3 x 3)

©Laura Putman, Bright Minds Engaged, 2023-present All rights reserved.

©Laura Putman, Bright Minds Engaged, 2023-present All rights reserved.

PUZZLE 16

Directions: Cut these puzzle pieces apart. Glue the piece with the answer on top of each matching problem on the previous page.

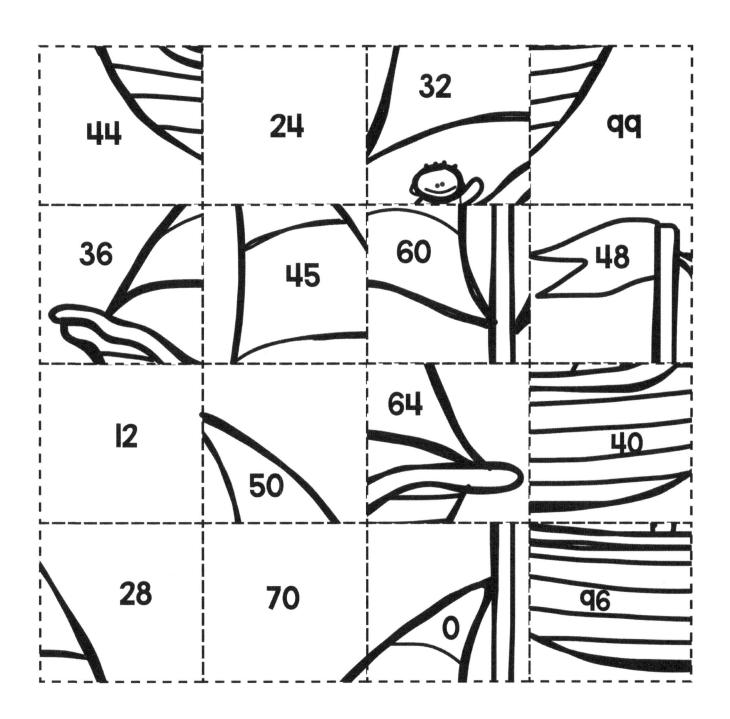

©Laura Putman, Bright Minds Engaged, 2023-present All rights reserved.

©Laura Putman, Bright Minds Engaged, 2023-present All rights reserved.

PUZZLE 17

Directions: Cut the puzzle pieces from the next page apart. To solve the puzzle, glue the piece with the matching answer on each problem.

5 × ? = 35 ? × 5 = 35	? × 8 = 40 8 × ? = 40	3 × 12 = ? 12 × 3 = ?
3 × ? = 12 ? × 3 = 12	? × 7 = 21 7 × ? = 21	9 × 10 = ? 10 × 9 = ?
? × 1 = 12 1 × ? = 12	12 × ? = 132 ? × 12 = 132	5 × ? = 30 ? × 5 = 30
? × 7 = 56 7 × ? = 56	2 × ? = 18 ? × 2 = 18	6 × ? = 60 ? × 6 = 60
9 × 7 = ? 7 × 9 = ?	4 × ? = 4 ? × 4 = 4	2 × ? = 0 ? × 2 = 0

©Laura Putman, Bright Minds Engaged, 2023-present All rights reserved.

©Laura Putman, Bright Minds Engaged, 2023-present All rights reserved.

PUZZLE 17 Directions: Cut these puzzle pieces apart. Glue the piece with the answer on top of each matching problem on the previous page.

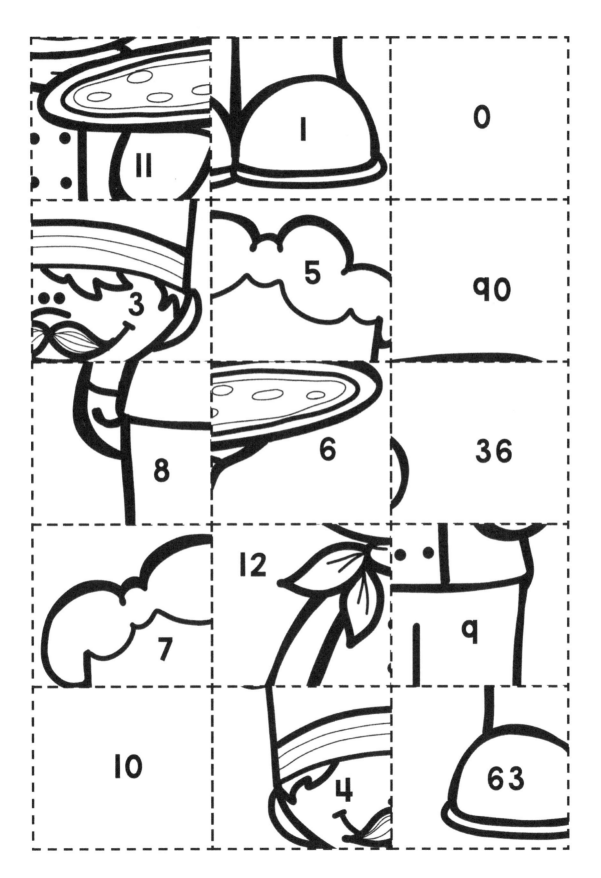

©Laura Putman, Bright Minds Engaged, 2023-present All rights reserved.

©Laura Putman, Bright Minds Engaged, 2023-present All rights reserved.

PUZZLE 18

Directions: Cut the puzzle pieces from the next page apart. To solve the puzzle, glue the piece with the matching answer on each problem.

6 × ? = 12 ? × 6 = 12	11 × 4 = ? 4 × 11 = ?	? × 9 = 0 9 × ? = 0	1 × ? = 3 ? × 1 = 3
8 × 2 = ? 2 × 8 = ?	? × 6 = 30 6 × ? = 30	? × 6 = 24 6 × ? = 24	8 × 9 = ? 9 × 8 = ?
10 × 3 = ? 3 × 10 = ?	7 × ? = 84 ? × 7 = 84	4 × 5 = ? 5 × 4 = ?	11 × 8 = ? 8 × 11 = ?
10 × ? = 110 ? × 10 = 110	4 × ? = 32 ? × 4 = 32	3 × 2 = ? 2 × 3 = ?	? × 5 = 45 5 × ? = 45

©Laura Putman, Bright Minds Engaged, 2023-present All rights reserved.

©Laura Putman, Bright Minds Engaged, 2023-present All rights reserved.

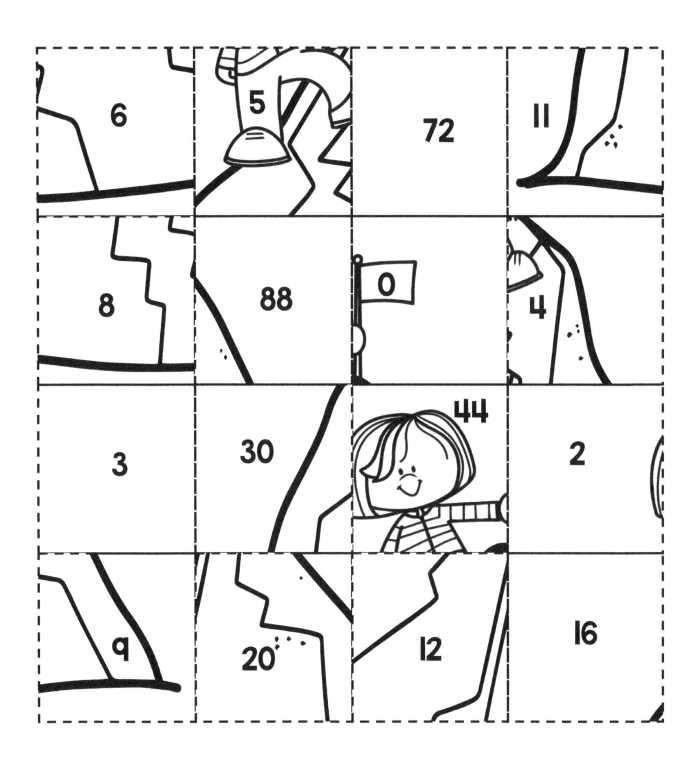

©Laura Putman, Bright Minds Engaged, 2023-present All rights reserved.

PUZZLE 19

Directions: Cut the puzzle pieces from the next page apart. To solve the puzzle, glue the piece with the matching answer on each problem.

4 × 8	9 × 5	11 × 6
___×___ + ___×___	___×___ + ___×___	___×___ + ___×___
____ + ____	____ + ____	____ + ____
= ____	= ____	= ____
7 × 2	10 × 3	5 × 12
___×___ + ___×___	___×___ + ___×___	___×___ + ___×___
____ + ____	____ + ____	____ + ____
= ____	= ____	= ____
12 × 12	7 × 8	6 × 9
___×___ + ___×___	___×___ + ___×___	___×___ + ___×___
____ + ____	____ + ____	____ + ____
= ____	= ____	= ____

©Laura Putman, Bright Minds Engaged, 2023-present All rights reserved.

©Laura Putman, Bright Minds Engaged, 2023-present All rights reserved.

PUZZLE 19

Directions: Cut these puzzle pieces apart. Glue the piece with the answer on top of each matching problem on the previous page.

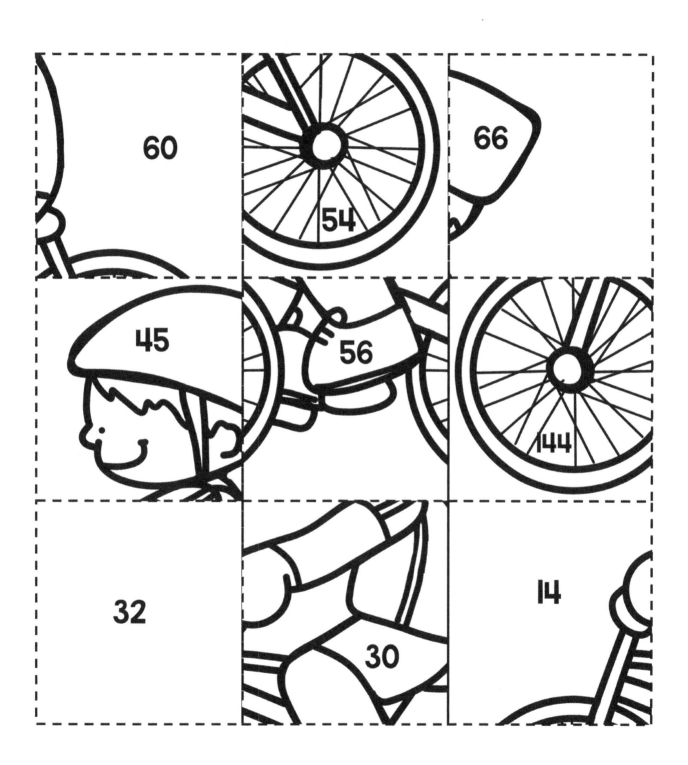

©Laura Putman, Bright Minds Engaged, 2023-present All rights reserved.

©Laura Putman, Bright Minds Engaged, 2023-present All rights reserved.

PUZZLE 20

Directions: Cut the puzzle pieces from the next page apart. To solve the puzzle, glue the piece with the matching answer on each problem.

9 × 9 ___x___ + ___x___ ___ + ___ = ___	**7 × 3** ___x___ + ___x___ ___ + ___ = ___	**10 × 7** ___x___ + ___x___ ___ + ___ = ___
12 × 6 ___x___ + ___x___ ___ + ___ = ___	**8 × 5** ___x___ + ___x___ ___ + ___ = ___	**11 × 4** ___x___ + ___x___ ___ + ___ = ___
3 × 9 ___x___ + ___x___ ___ + ___ = ___	**8 × 8** ___x___ + ___x___ ___ + ___ = ___	**11 × 11** ___x___ + ___x___ ___ + ___ = ___
10 × 6 ___x___ + ___x___ ___ + ___ = ___	**9 × 2** ___x___ + ___x___ ___ + ___ = ___	**12 × 7** ___x___ + ___x___ ___ + ___ = ___

©Laura Putman, Bright Minds Engaged, 2023-present All rights reserved.

©Laura Putman, Bright Minds Engaged, 2023-present All rights reserved.

PUZZLE 20

Directions: Cut these puzzle pieces apart. Glue the piece with the answer on top of each matching problem on the previous page.

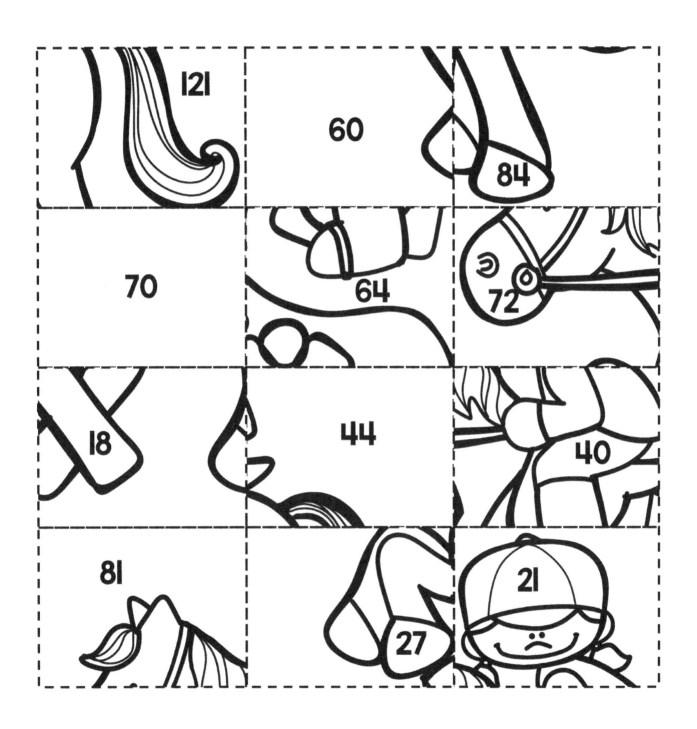

©Laura Putman, Bright Minds Engaged, 2023-present All rights reserved.

©Laura Putman, Bright Minds Engaged, 2023-present All rights reserved.

PUZZLE 21

Directions: Cut the puzzle pieces from the next page apart. To solve the puzzle, glue the piece with the matching answer on each problem.

8 × 50	20 × 3	10 × 12
10 × 11	7 × 70	80 × 4
6 × 90	50 × 4	9 × 30
80 × 8	8 × 90	2 × 20
40 × 4	60 × 3	12 × 50

©Laura Putman, Bright Minds Engaged, 2023-present All rights reserved.

©Laura Putman, Bright Minds Engaged, 2023-present All rights reserved.

PUZZLE 21

Directions: Cut these puzzle pieces apart. Glue the piece with the answer on top of each matching problem on the previous page.

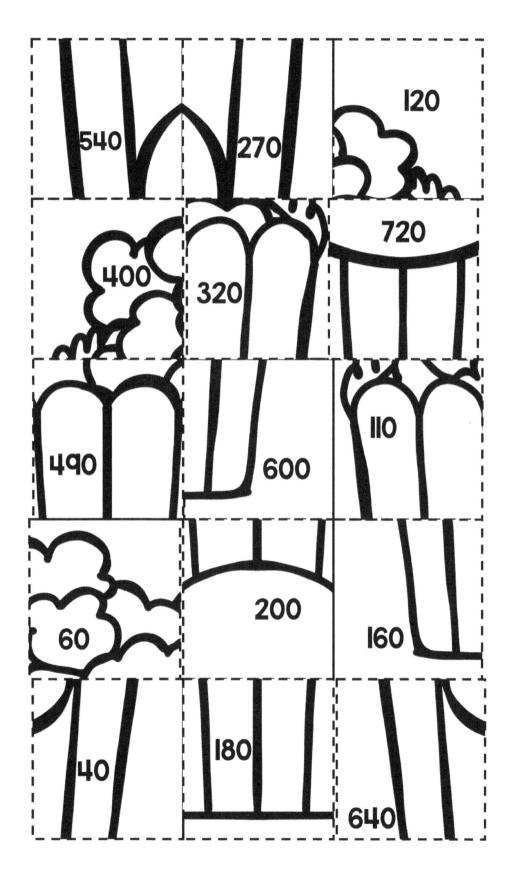

©Laura Putman, Bright Minds Engaged, 2023-present All rights reserved.

©Laura Putman, Bright Minds Engaged, 2023-present All rights reserved.

PUZZLE 22

Directions: Cut the puzzle pieces from the next page apart. To solve the puzzle, glue the piece with the matching answer on each problem.

10 x 80	30 x 7	12 x 40	9 x 60
5 x 60	20 x 3	40 x 2	11 x 60
40 x 7	10 x 90	8 x 90	50 x 12
80 x 3	70 x 6	8 x 80	11 x 20

©Laura Putman, Bright Minds Engaged, 2023-present All rights reserved.

©Laura Putman, Bright Minds Engaged, 2023-present All rights reserved.

PUZZLE 22

Directions: Cut these puzzle pieces apart. Glue the piece with the answer on top of each matching problem on the previous page.

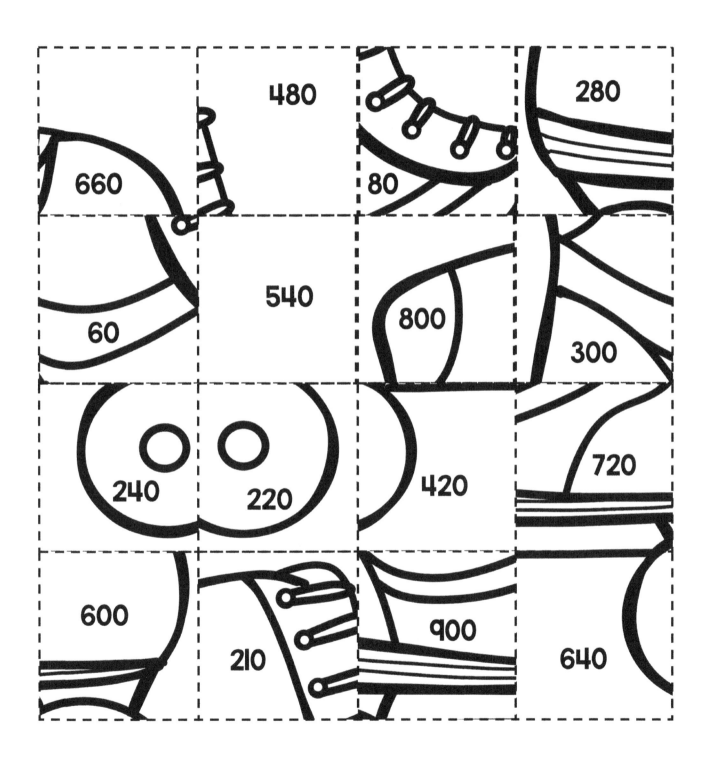

©Laura Putman, Bright Minds Engaged, 2023-present All rights reserved.

PUZZLE 23

Directions: Cut the puzzle pieces from the next page apart. To solve the puzzle, glue the piece with the matching answer on each problem.

? ÷ 9 = 9	32 ÷ ? = 4	? ÷ 8 = 8	36 ÷ 9 = ?
? ÷ 7 = 2	? ÷ 7 = 9	99 ÷ ? = 9	60 ÷ 10 = ?
45 ÷ ? = 9	144 ÷ ? = 12	? ÷ 5 = 12	? ÷ 3 = 9
? ÷ 7 = 11	90 ÷ ? = 9	? ÷ 3 = 11	? ÷ 11 = 11
? ÷ 12 = 6	? ÷ 8 = 8	12 ÷ 6 = ?	5 ÷ ? = 5

©Laura Putman, Bright Minds Engaged, 2023-present All rights reserved.

©Laura Putman, Bright Minds Engaged, 2023-present All rights reserved.

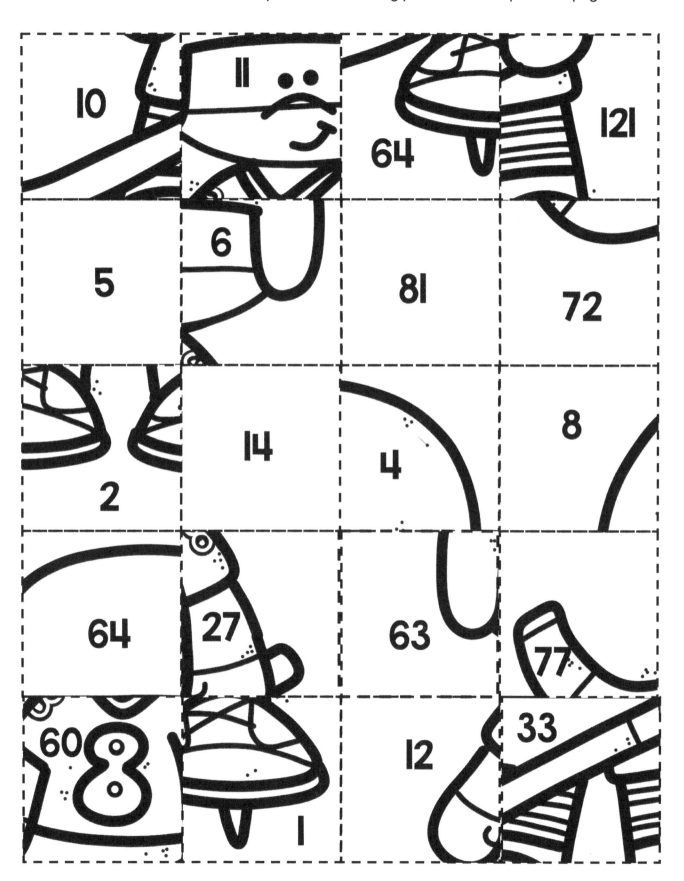

©Laura Putman, Bright Minds Engaged, 2023-present All rights reserved.

PUZZLE 24

Directions: Cut the puzzle pieces from the next page apart. To solve the puzzle, glue the piece with the matching answer on each problem.

18 ÷ ? = 2	? ÷ 7 = 8	? ÷ 3 = 7
0 ÷ 12 = ?	? ÷ 5 = 5	11 ÷ ? = 11
? ÷ 2 = 2	9 ÷ 3 = ?	16 ÷ 8 = ?
? ÷ 4 = 8	? ÷ 8 = 6	54 ÷ ? = 9
? ÷ 7 = 2	80 ÷ ? = 8	30 ÷ ? = 6

©Laura Putman, Bright Minds Engaged, 2023-present All rights reserved.

©Laura Putman, Bright Minds Engaged, 2023-present All rights reserved.

Directions: Cut these puzzle pieces apart. Glue the piece with the answer on top of each matching problem on the previous page.

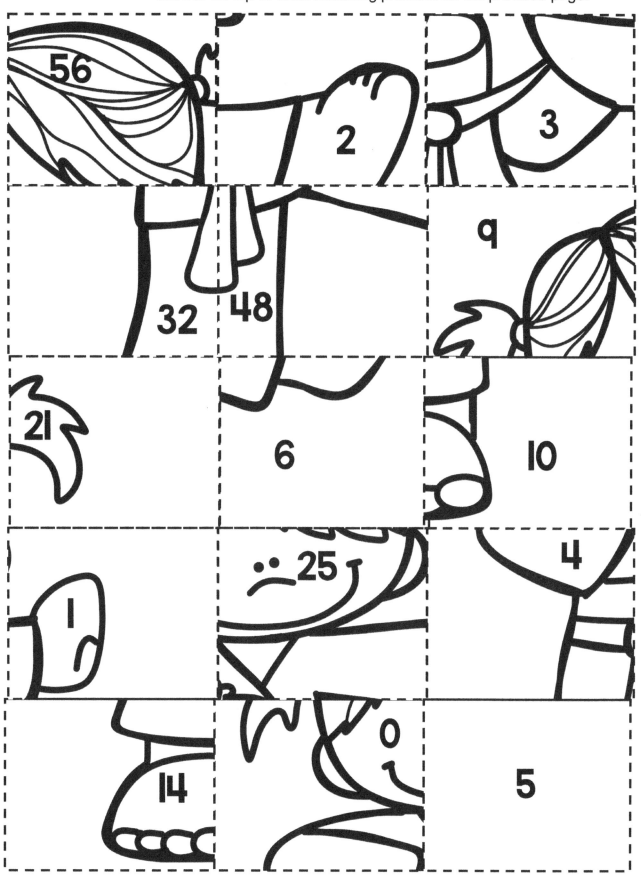

©Laura Putman, Bright Minds Engaged, 2023-present All rights reserved.

©Laura Putman, Bright Minds Engaged, 2023-present All rights reserved.

PUZZLE 25

Directions: Cut the puzzle pieces from the next page apart. To solve the puzzle, glue the piece with the matching answer on each problem.

6x5	49÷7	72÷8	4x4	3x11
100÷10	5x12	6x7	3x4	40÷8
6x6	2x12	0÷4	8x7	21÷7
10x11	9x9	6x7	16÷8	9x2
80÷10	121÷11	12x6	48÷8	20÷5

©Laura Putman, Bright Minds Engaged, 2023-present All rights reserved.

©Laura Putman, Bright Minds Engaged, 2023-present All rights reserved.

PUZZLE 25

Directions: Cut these puzzle pieces apart. Glue the piece with the answer on top of each matching problem on the previous page.

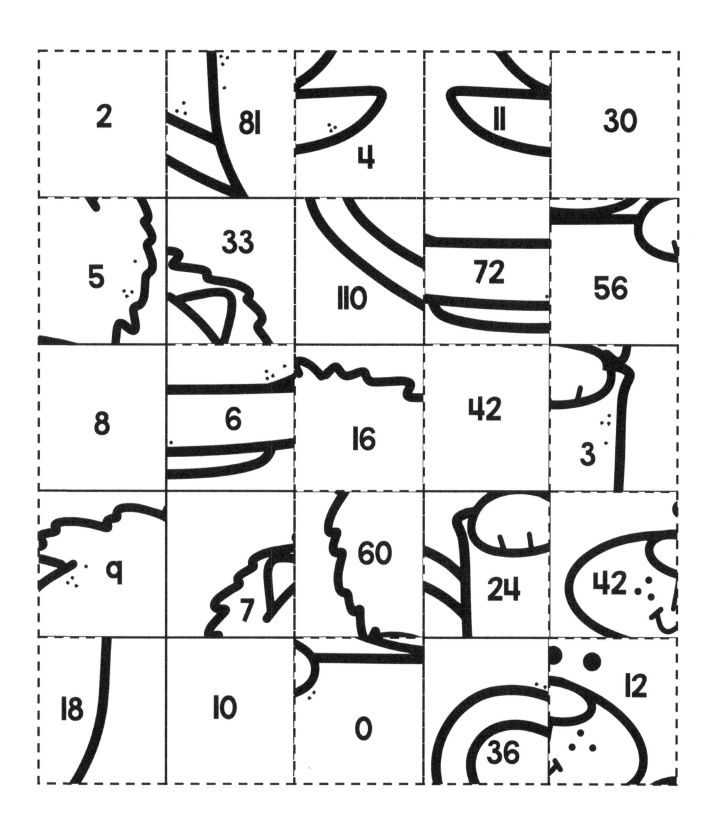

©Laura Putman, Bright Minds Engaged, 2023-present All rights reserved.

©Laura Putman, Bright Minds Engaged, 2023-present All rights reserved.

PUZZLE 26

Directions: Cut the puzzle pieces from the next page apart. To solve the puzzle, glue the piece with the matching answer on each problem.

77÷11	5×3	108÷12	6×8
4×6	16÷4	7×7	25÷5
64÷8	100÷10	121÷11	72÷6
9÷3	0÷10	9×9	42÷7
10÷5	8×4	1÷1	4×5

©Laura Putman, Bright Minds Engaged, 2023-present All rights reserved.

©Laura Putman, Bright Minds Engaged, 2023-present All rights reserved.

PUZZLE 26

Directions: Cut these puzzle pieces apart. Glue the piece with the answer on top of each matching problem on the previous page.

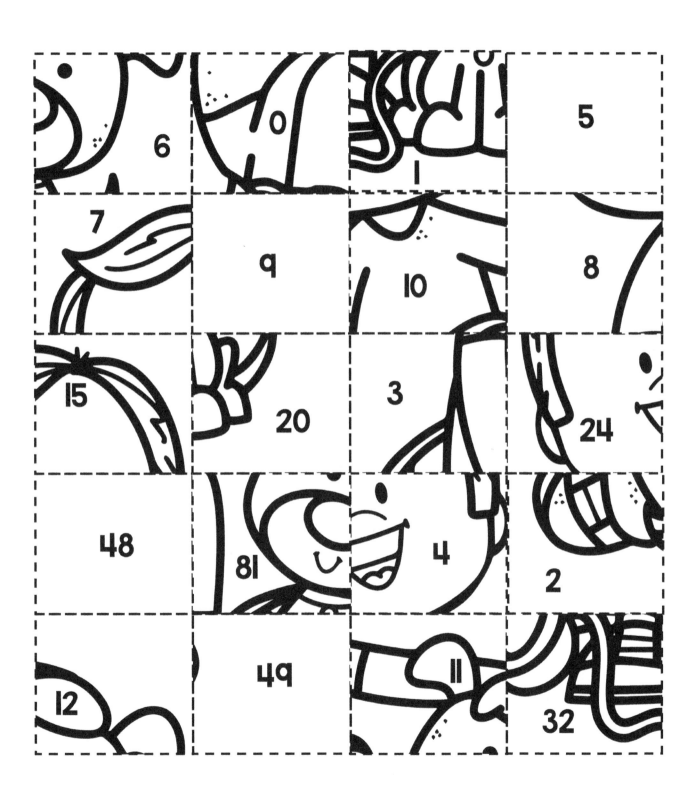

©Laura Putman, Bright Minds Engaged, 2023-present All rights reserved.

©Laura Putman, Bright Minds Engaged, 2023-present All rights reserved.

PUZZLE 27

Directions: Cut the puzzle pieces from the next page apart. To solve the puzzle, glue the piece with the matching answer on each problem.

3 × 2 = ? 2 × 3 = ? ? ÷ 3 = 2 ? ÷ 2 = 3	? × 9 = 45 9 × ? = 45 45 ÷ ? = 9 45 ÷ 9 = ?	? × 5 = 55 5 × ? = 55 55 ÷ 5 = ? 55 ÷ ? = 5	7 × 8 = ? 8 × 7 = ? ? ÷ 8 = 7 ? ÷ 7 = 8
9 × 4 = ? 4 × 9 = ? ? ÷ 9 = 4 ? ÷ 4 = 9	3 × 1 = ? 1 × 3 = ? ? ÷ 1 = 3 ? ÷ 3 = 1	? × 8 = 0 8 × ? = 0 ? ÷ 8 = 0	? × 3 = 12 3 × ? = 12 12 ÷ 3 = ? 12 ÷ ? = 3
6 × 2 = ? 2 × 6 = ? ? ÷ 2 = 6 ? ÷ 6 = 2	7 × 12 = ? 12 × 7 = ? ? ÷ 7 = 12 ? ÷ 12 = 7	9 × 10 = ? 10 × 9 = ? ? ÷ 9 = 10 ? ÷ 10 = 9	4 × 4 = ? ? ÷ 4 = 4
10 × 10 = ? ? ÷ 10 = 10	5 × 12 = ? 12 × 5 = ? ? ÷ 12 = 5 ? ÷ 5 = 12	? × 5 = 5 5 × ? = 5 5 ÷ ? = 5 5 ÷ 5 = ?	? × 2 = 4 2 × ? = 4 4 ÷ ? = 2 4 ÷ 2 = ?

PUZZLE 27

Directions: Cut these puzzle pieces apart. Glue the piece with the answer on top of each matching problem on the previous page.

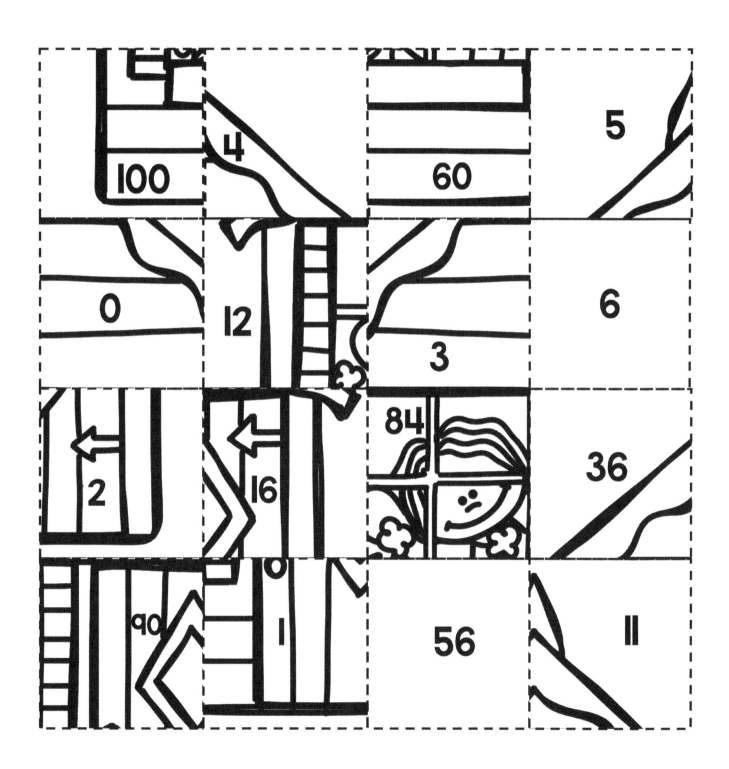

©Laura Putman, Bright Minds Engaged, 2023-present All rights reserved.

©Laura Putman, Bright Minds Engaged, 2023-present All rights reserved.

PUZZLE 28

Directions: Cut the puzzle pieces from the next page apart. To solve the puzzle, glue the piece with the matching answer on each problem.

8 × 8 = ? ? ÷ 8 = 8	? × 7 = 63 7 × ? = 63 63 ÷ ? = 7 63 ÷ 7 = ?	? × 4 = 20 4 × ? = 20 20 ÷ 4 = ? 20 ÷ ? = 4	3 × 6 = ? 6 × 3 = ? ? ÷ 6 = 3 ? ÷ 3 = 6
5 × 12 = ? 12 × 5 = ? ? ÷ 5 = 12 ? ÷ 12 = 5	2 × ? = 4 ? × 2 = 4 4 ÷ ? = 2 4 ÷ 2 = ?	? × 10 = 110 10 × ? = 110 110 ÷ 10 = ? 110 ÷ ? = 10	? × 8 = 56 8 × ? = 56 56 ÷ 8 = ? 56 ÷ ? = 8
6 × 4 = ? 4 × 6 = ? ? ÷ 4 = 6 ? ÷ 6 = 4	9 × 9 = ? ? ÷ 9 = 9	4 × ? = 12 ? × 4 = 12 12 ÷ ? = 4 12 ÷ 4 = ?	7 × ? = 42 ? × 7 = 42 42 ÷ ? = 7 42 ÷ 7 = ?
1 × 8 = ? 8 × 1 = ? ? ÷ 1 = 8 8 ÷ ? = 1	11 × 2 = ? 2 × 11 = ? ? ÷ 2 = 11 ? ÷ 11 = 2	? × 8 = 96 8 × ? = 96 96 ÷ ? = 8 96 ÷ 8 = ?	3 × 5 = ? 5 × 3 = ? ? ÷ 3 = 5 ? ÷ 5 = 3

©Laura Putman, Bright Minds Engaged, 2023-present All rights reserved.

©Laura Putman, Bright Minds Engaged, 2023-present All rights reserved.

PUZZLE 28

Directions: Cut these puzzle pieces apart. Glue the piece with the answer on top of each matching problem on the previous page.

©Laura Putman, Bright Minds Engaged, 2023-present All rights reserved.

©Laura Putman, Bright Minds Engaged, 2023-present All rights reserved.

PUZZLE 29

Directions: Cut the puzzle pieces from the next page apart. To solve the puzzle, glue the piece with the matching polygon name on each shape.

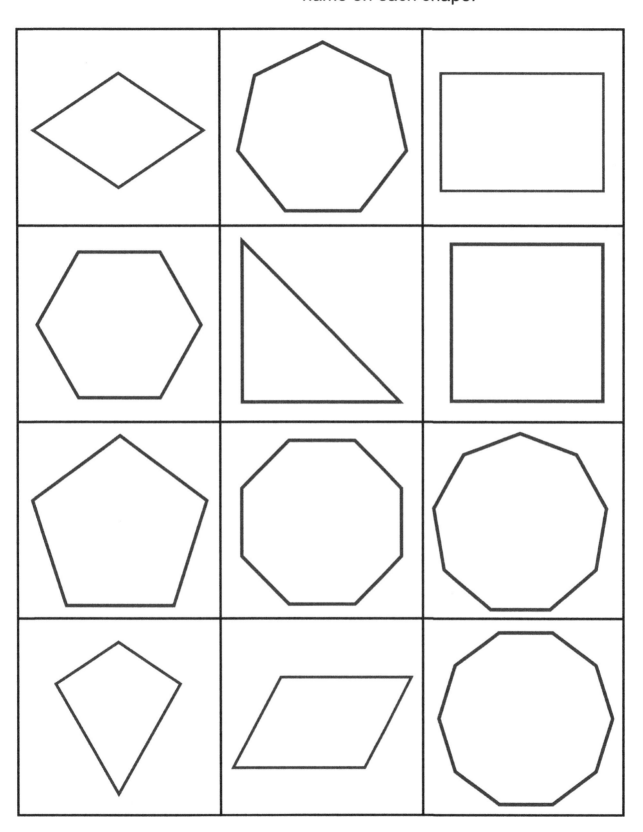

©Laura Putman, Bright Minds Engaged, 2023-present All rights reserved.

©Laura Putman, Bright Minds Engaged, 2023-present All rights reserved.

PUZZLE 29 Directions: Cut these puzzle pieces apart. Glue the piece with the polygon name on top of each matching polygon on the previous page.

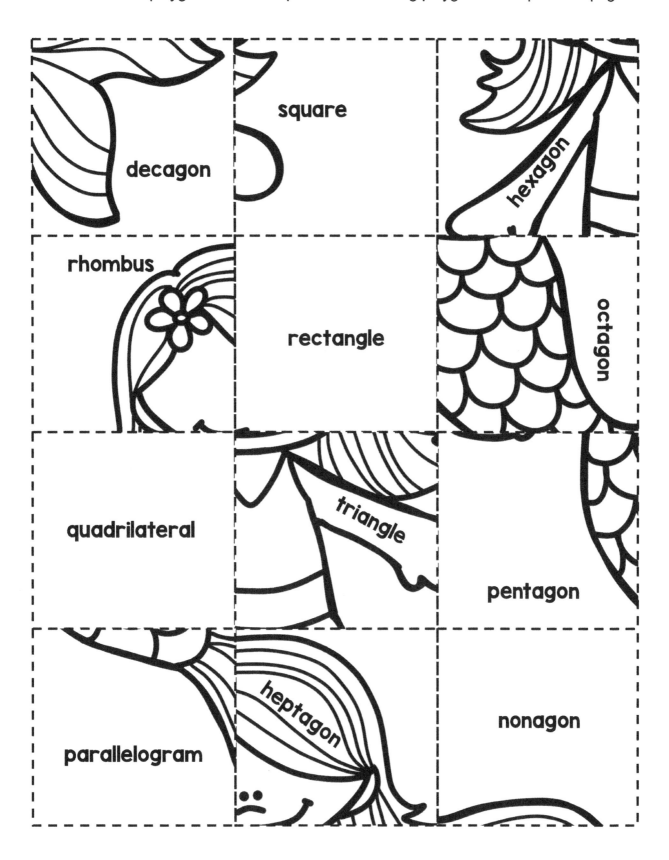

©Laura Putman, Bright Minds Engaged, 2023-present All rights reserved.

©Laura Putman, Bright Minds Engaged, 2023-present All rights reserved.

PUZZLE 30

Directions: Cut the puzzle pieces from the next page apart. To solve the puzzle, glue the piece with the matching polygon name on each shape.

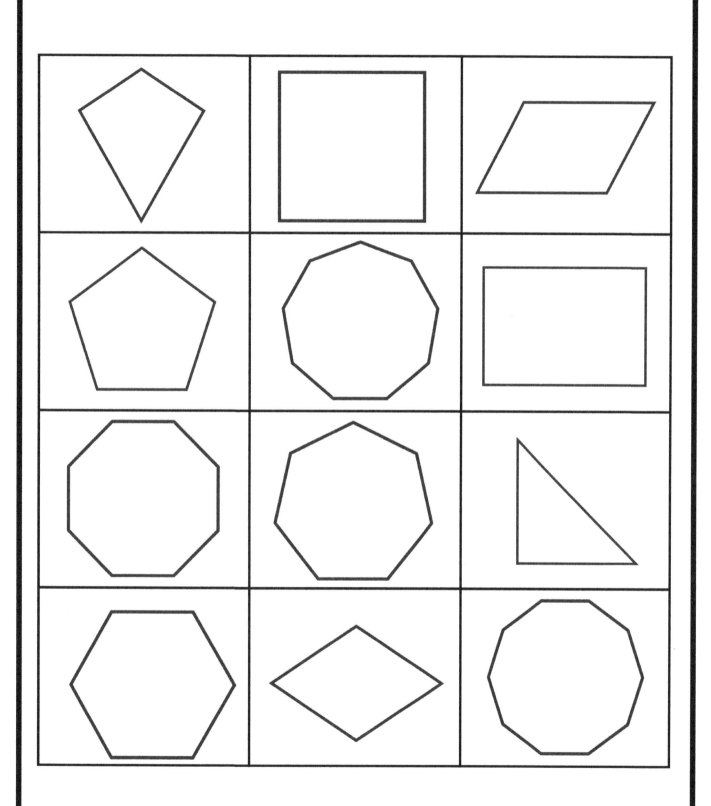

©Laura Putman, Bright Minds Engaged, 2023-present All rights reserved.

©Laura Putman, Bright Minds Engaged, 2023-present All rights reserved.

Directions: Cut these puzzle pieces apart. Glue the piece with the polygon name on top of each matching polygon on the previous page.

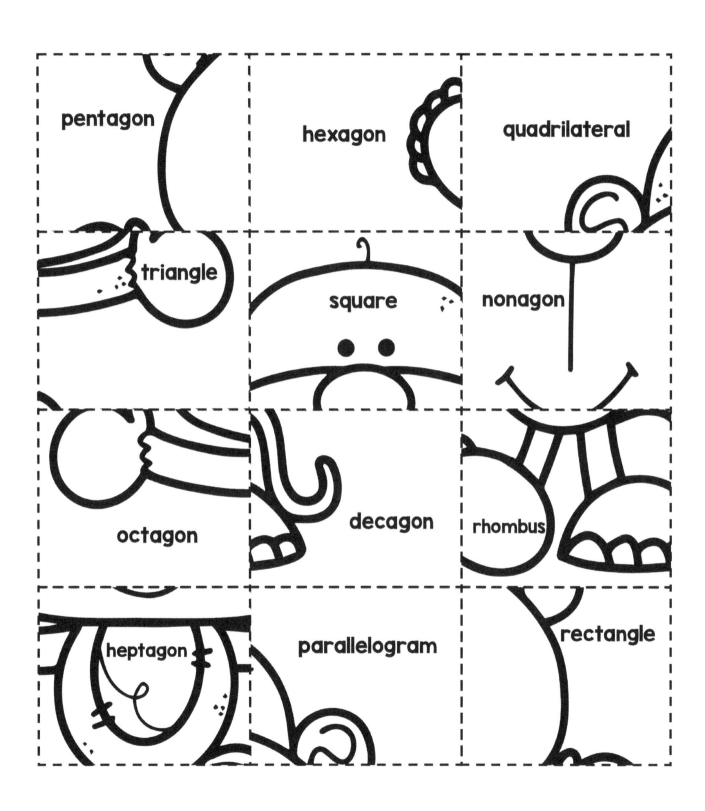

©Laura Putman, Bright Minds Engaged, 2023-present All rights reserved.

PUZZLE 31

Directions: Cut the puzzle pieces from the next page apart. To solve the puzzle, glue the piece with the matching area on each rectangle.

5 in × 2 in	12 cm × 7 cm	3 cm × 8 cm	9 in × 9 in
11 cm × 11 cm	6 in × 1 in	12 in × 9 in	10 cm × 4 cm
2 in × 4 in	6 in × 6 in	1 cm × 8 cm	11 in × 3 in
7 cm × 7 cm	10 in × 6 in	3 in × 5 in	2 cm × 8 cm

©Laura Putman, Bright Minds Engaged, 2023-present All rights reserved.

©Laura Putman, Bright Minds Engaged, 2023-present All rights reserved.

PUZZLE 31

Directions: Cut these puzzle pieces apart. Glue the piece with the area on top of each matching rectangle on the previous page.

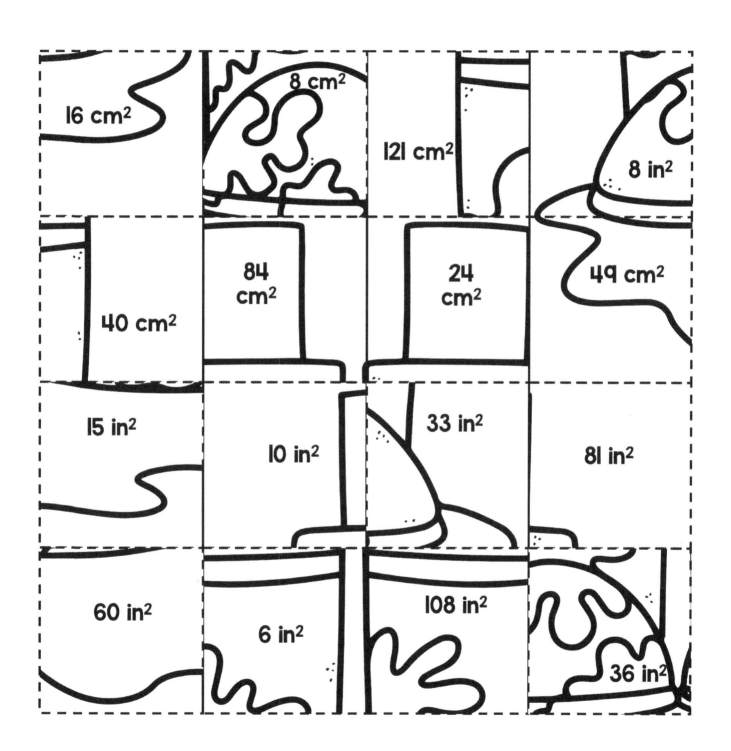

©Laura Putman, Bright Minds Engaged, 2023-present All rights reserved.

©Laura Putman, Bright Minds Engaged, 2023-present All rights reserved.

PUZZLE 32

Directions: Cut the puzzle pieces from the next page apart. To solve the puzzle, glue the piece with the matching area on each rectangle.

6 in × 12 in	1 cm × 7 cm	7 in × 11 in
9 cm × 9 cm	10 in × 5 in	12 cm × 12 cm
2 cm × 7 cm	6 in × 9 in	4 in × 4 in
3 in × 12 in	4 cm × 7 cm	6 in × 3 in

©Laura Putman, Bright Minds Engaged, 2023-present All rights reserved.

©Laura Putman, Bright Minds Engaged, 2023-present All rights reserved.

PUZZLE 32

Directions: Cut these puzzle pieces apart. Glue the piece with the area on top of each matching rectangle on the previous page.

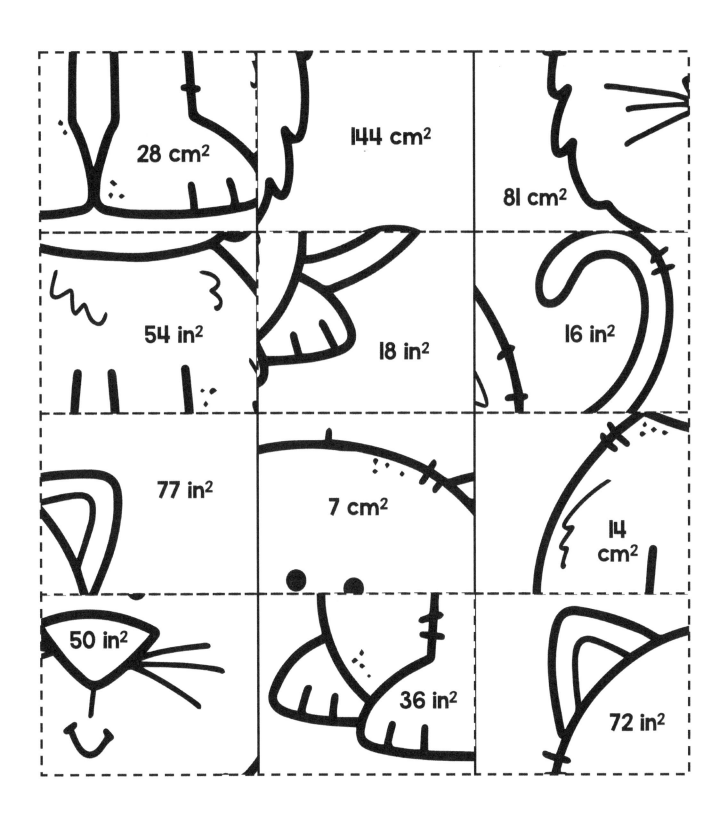

©Laura Putman, Bright Minds Engaged, 2023-present All rights reserved.

©Laura Putman, Bright Minds Engaged, 2023-present All rights reserved.

PUZZLE 33

Directions: Cut the puzzle pieces from the next page apart. To solve the puzzle, glue the piece with the matching perimeter on each shape.

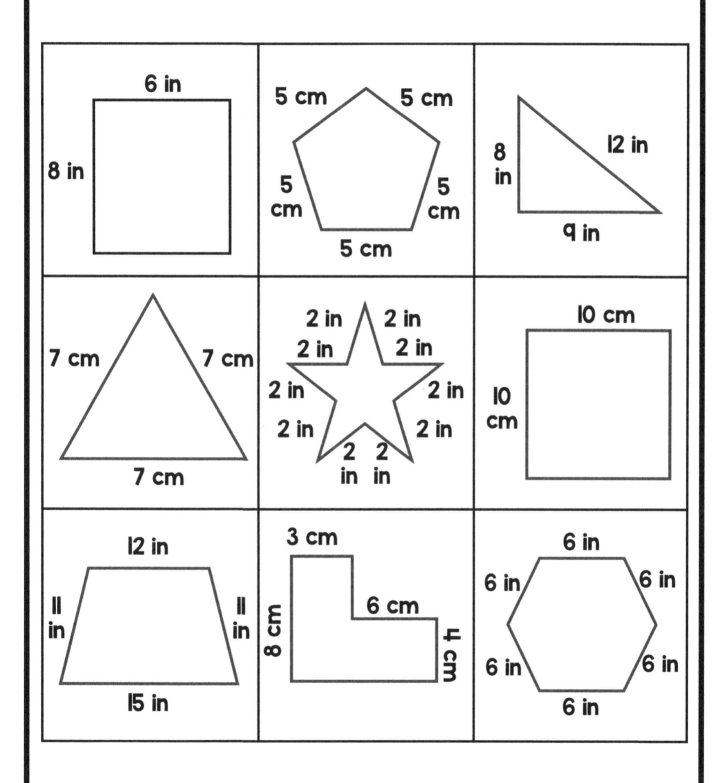

©Laura Putman, Bright Minds Engaged, 2023-present All rights reserved.

PUZZLE 33

Directions: Cut these puzzle pieces apart. Glue the piece with the correct perimeter on top of each matching shape on the previous page.

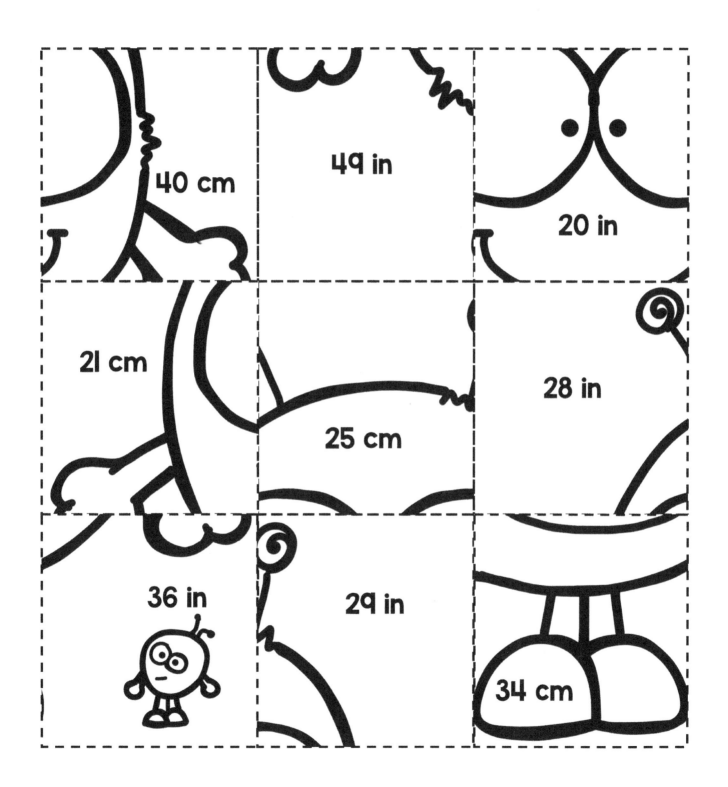

©Laura Putman, Bright Minds Engaged, 2023-present All rights reserved.

©Laura Putman, Bright Minds Engaged, 2023-present All rights reserved.

PUZZLE 34

Directions: Cut the puzzle pieces from the next page apart. To solve the puzzle, glue the piece with the matching perimeter on each shape.

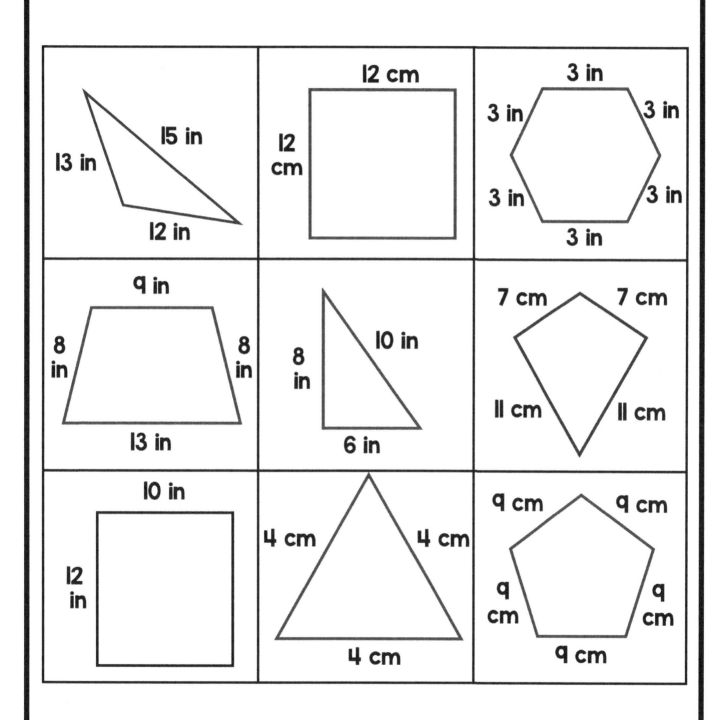

©Laura Putman, Bright Minds Engaged, 2023-present All rights reserved.

©Laura Putman, Bright Minds Engaged, 2023-present All rights reserved.

PUZZLE 34

Directions: Cut these puzzle pieces apart. Glue the piece with the correct perimeter on top of each matching shape on the previous page.

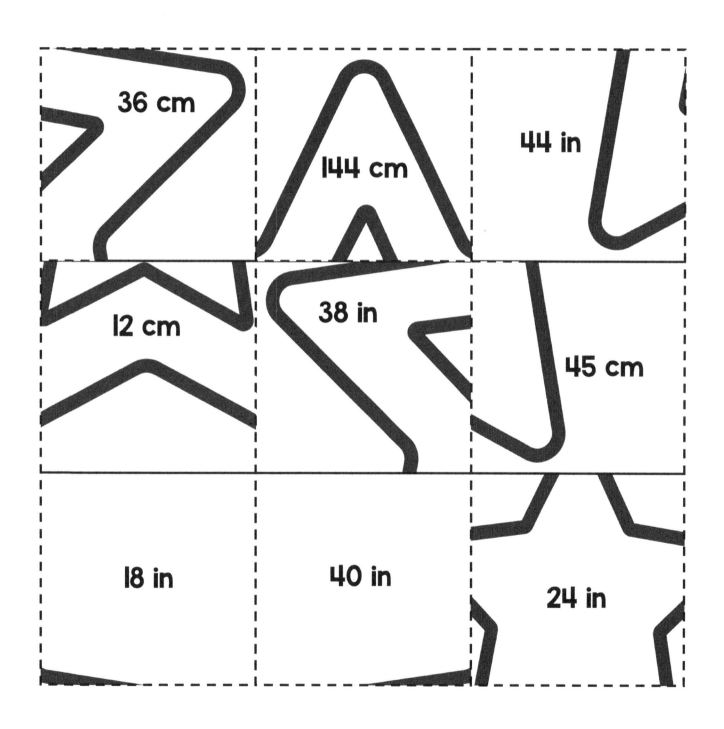

©Laura Putman, Bright Minds Engaged, 2023-present All rights reserved.

©Laura Putman, Bright Minds Engaged, 2023-present All rights reserved.

PUZZLE 35

Directions: Cut the puzzle pieces from the next page apart. To solve the puzzle, glue the piece with the matching fraction name on each picture.

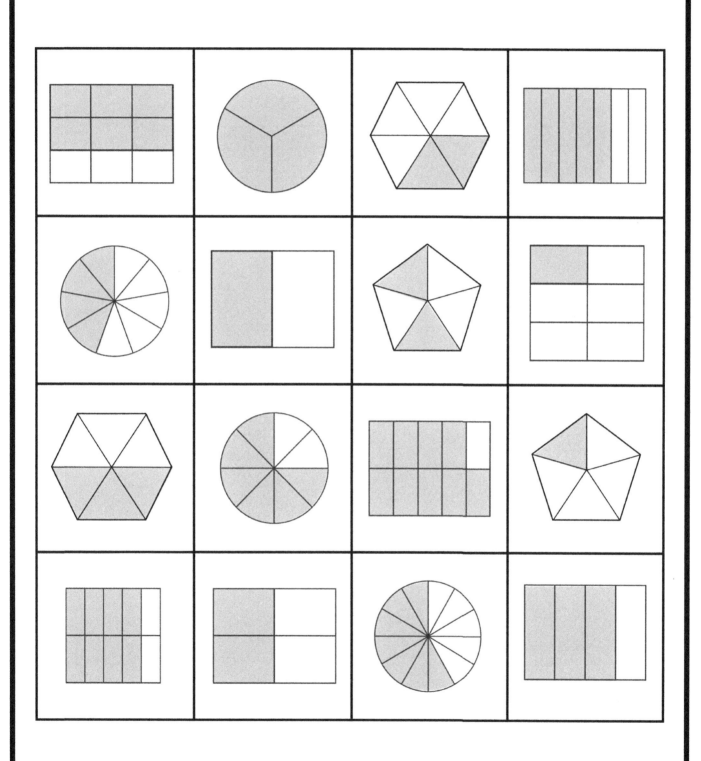

©Laura Putman, Bright Minds Engaged, 2023-present All rights reserved.

©Laura Putman, Bright Minds Engaged, 2023-present All rights reserved.

PUZZLE 35

Directions: Cut these puzzle pieces apart. Glue the piece with the correct fraction name on top of each matching fraction on the previous page.

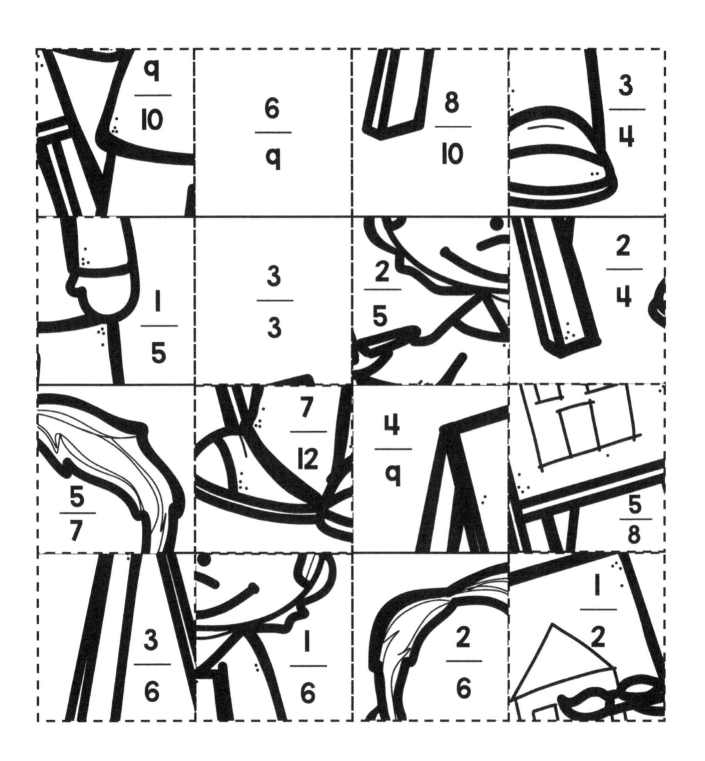

©Laura Putman, Bright Minds Engaged, 2023-present All rights reserved.

©Laura Putman, Bright Minds Engaged, 2023-present All rights reserved.

PUZZLE 36

Directions: Cut the puzzle pieces from the next page apart. To solve the puzzle, glue the piece with the matching fraction name on each picture.

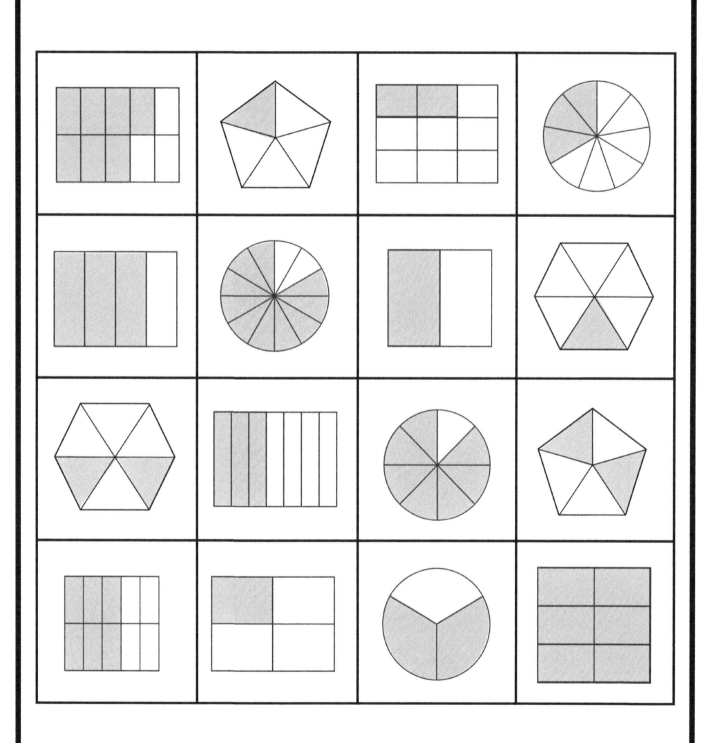

©Laura Putman, Bright Minds Engaged, 2023-present All rights reserved.

©Laura Putman, Bright Minds Engaged, 2023-present All rights reserved.

PUZZLE 36

Directions: Cut these puzzle pieces apart. Glue the piece with the correct fraction name on top of each matching fraction on the previous page.

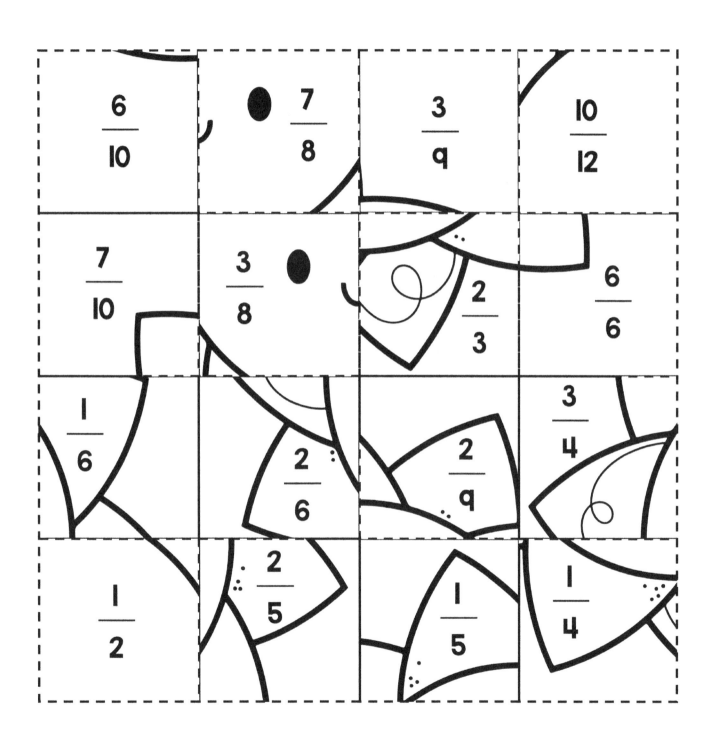

©Laura Putman, Bright Minds Engaged, 2023-present All rights reserved.

©Laura Putman, Bright Minds Engaged, 2023-present All rights reserved.

PUZZLE 37

Directions: Cut the puzzle pieces from the next page apart. To solve the puzzle, glue the piece with the matching fraction name on each number line.

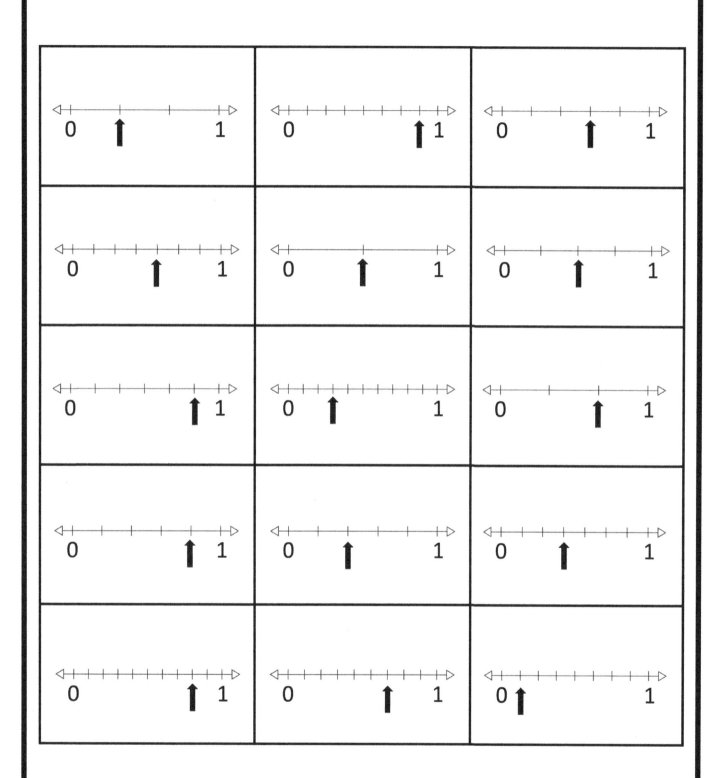

©Laura Putman, Bright Minds Engaged, 2023-present All rights reserved.

©Laura Putman, Bright Minds Engaged, 2023-present All rights reserved.

PUZZLE 37

Directions: Cut these puzzle pieces apart. Glue the piece with the correct fraction name on top of each matching number line on the previous page.

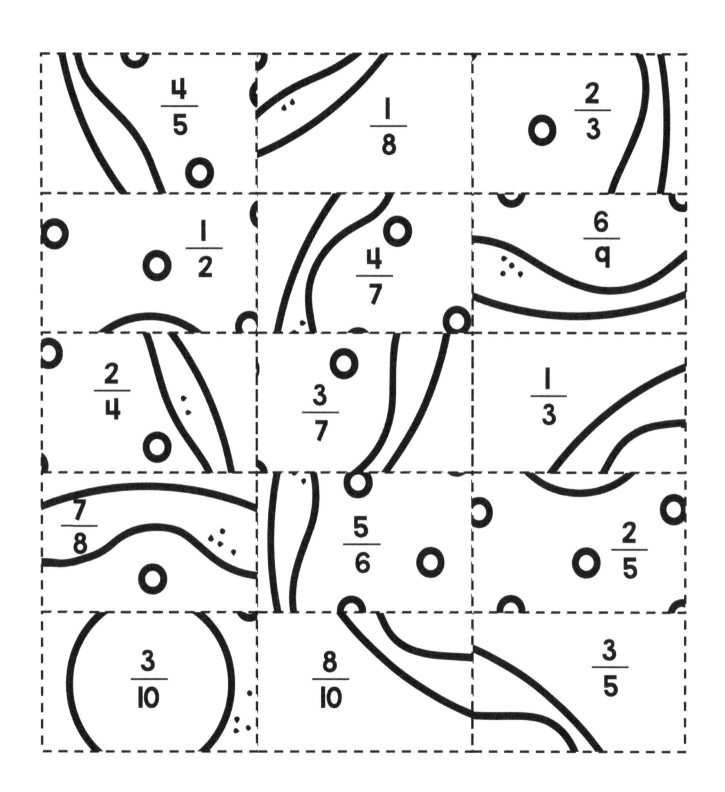

©Laura Putman, Bright Minds Engaged, 2023-present All rights reserved.

PUZZLE 38

Directions: Cut the puzzle pieces from the next page apart. To solve the puzzle, glue the piece with the matching fraction name on each number line.

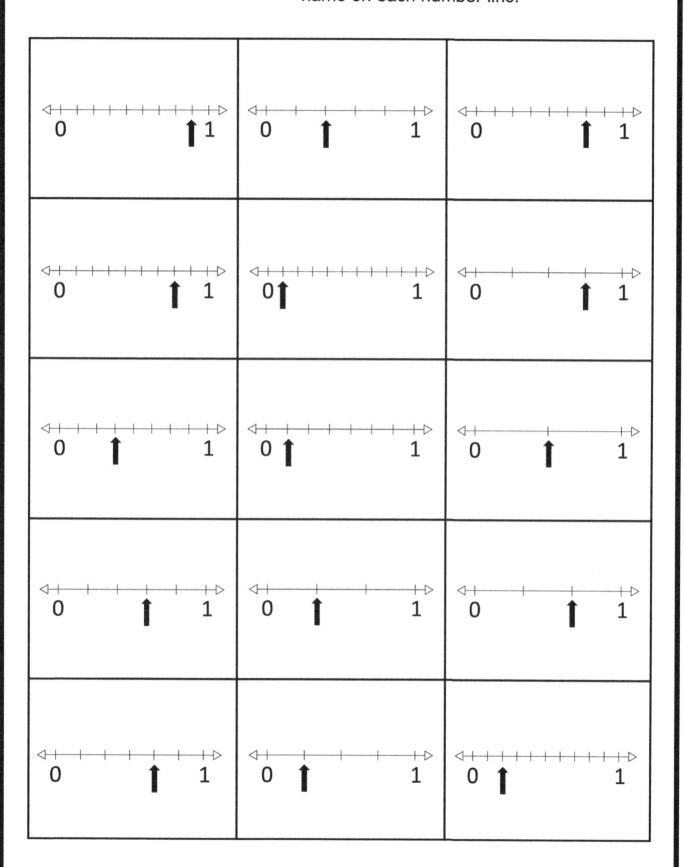

©Laura Putman, Bright Minds Engaged, 2023-present All rights reserved.

©Laura Putman, Bright Minds Engaged, 2023-present All rights reserved.

PUZZLE 38

Directions: Cut these puzzle pieces apart. Glue the piece with the correct fraction name on top of each matching number line on the previous page.

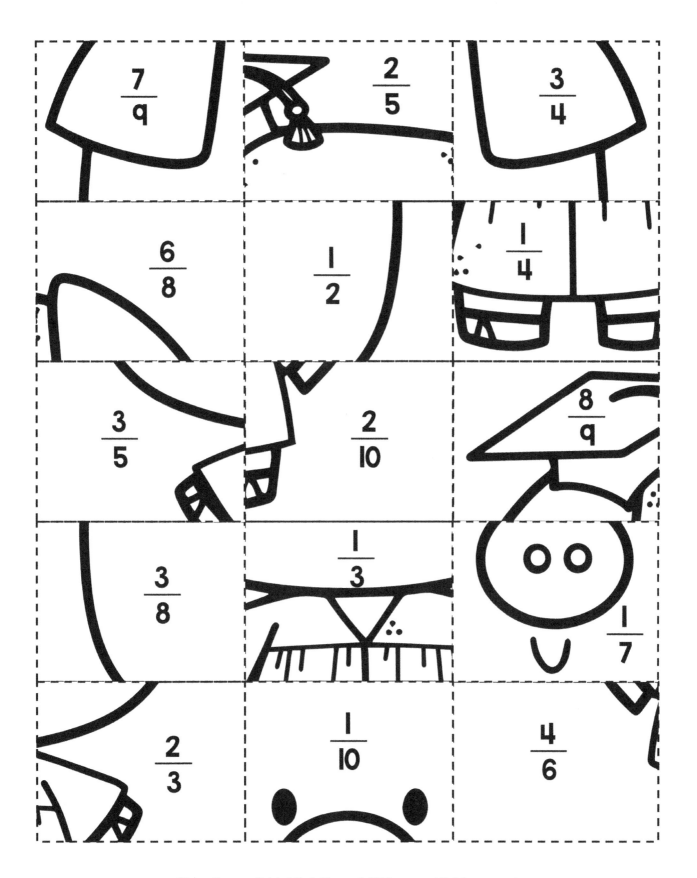

©Laura Putman, Bright Minds Engaged, 2023-present All rights reserved.

PUZZLE 39

Directions: Cut the puzzle pieces from the next page apart. To solve the puzzle, glue the piece with the equivalent fraction on each fraction.

$\dfrac{1}{2}$	$\dfrac{2}{3}$	$\dfrac{4}{5}$
$\dfrac{1}{3}$	$\dfrac{2}{8}$	$\dfrac{6}{10}$
$\dfrac{6}{8}$	$\dfrac{1}{5}$	$\dfrac{5}{6}$
$\dfrac{4}{4}$	$\dfrac{10}{16}$	$\dfrac{2}{12}$

©Laura Putman, Bright Minds Engaged, 2023-present All rights reserved.

©Laura Putman, Bright Minds Engaged, 2023-present All rights reserved.

PUZZLE 39

Directons: Cut these puzzle pieces apart. Glue the piece with the equivalent fraction on top of each fraction on the previous page.

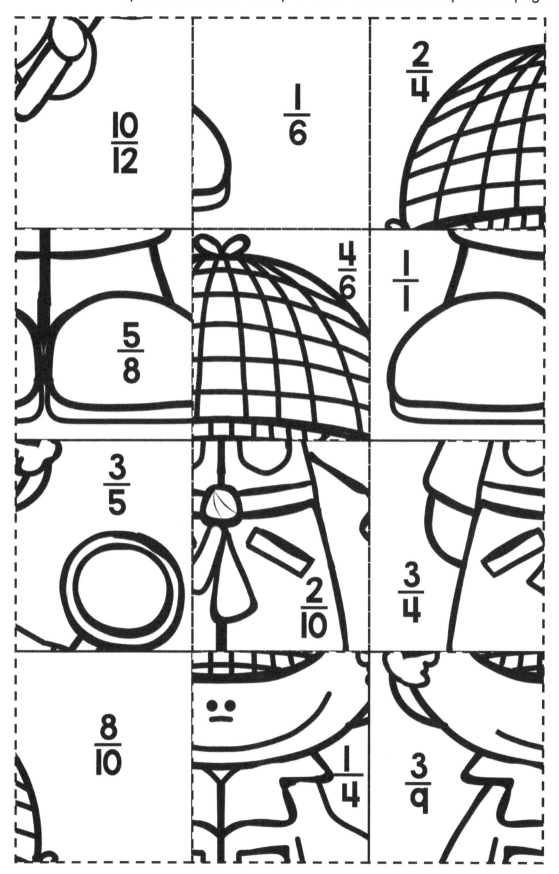

©Laura Putman, Bright Minds Engaged, 2023-present All rights reserved.

©Laura Putman, Bright Minds Engaged, 2023-present All rights reserved.

PUZZLE 40

Directions: Cut the puzzle pieces from the next page apart. To solve the puzzle, glue the piece with the equivalent fraction on each fraction.

$\frac{2}{8}$	$\frac{4}{8}$	$\frac{1}{5}$
$\frac{2}{2}$	$\frac{3}{4}$	$\frac{6}{10}$
$\frac{3}{9}$	$\frac{4}{6}$	$\frac{4}{5}$
$\frac{3}{4}$	$\frac{2}{5}$	$\frac{5}{6}$

©Laura Putman, Bright Minds Engaged, 2023-present All rights reserved.

PUZZLE 40

Directions: Cut these puzzle pieces apart. Glue the piece with the equivalent fraction on top of each fraction on the previous page.

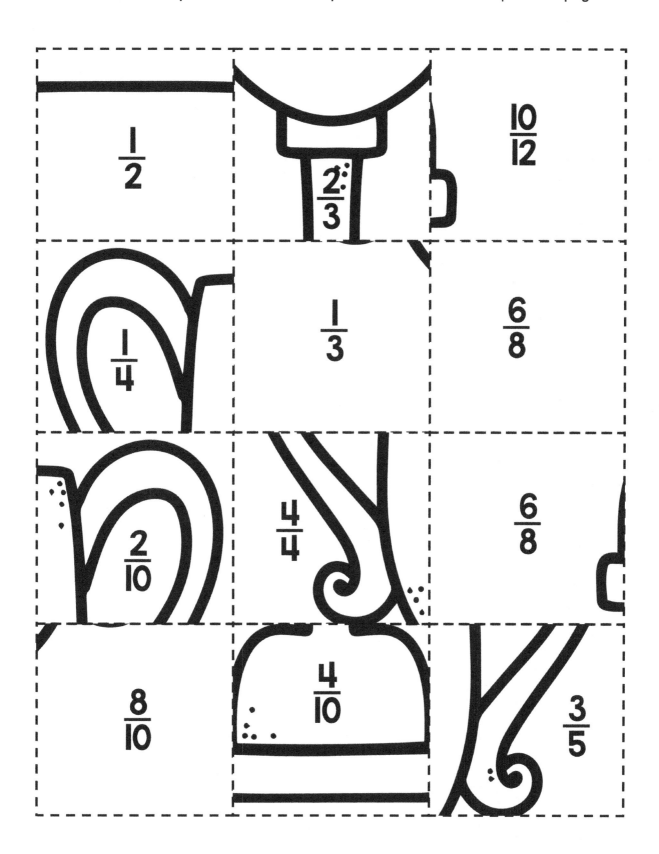

©Laura Putman, Bright Minds Engaged, 2023-present All rights reserved.

©Laura Putman, Bright Minds Engaged, 2023-present All rights reserved.

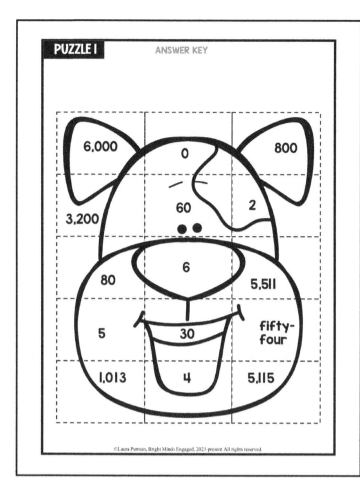
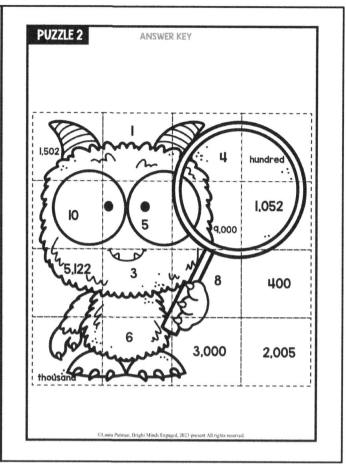
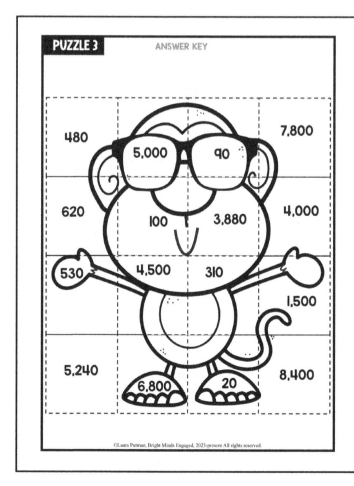
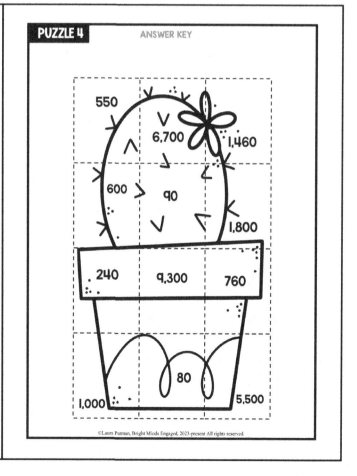

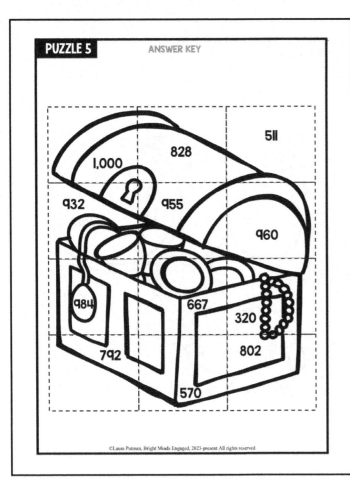
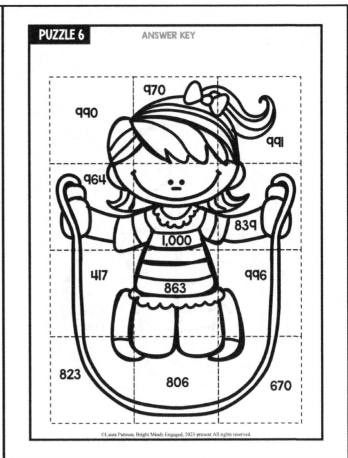
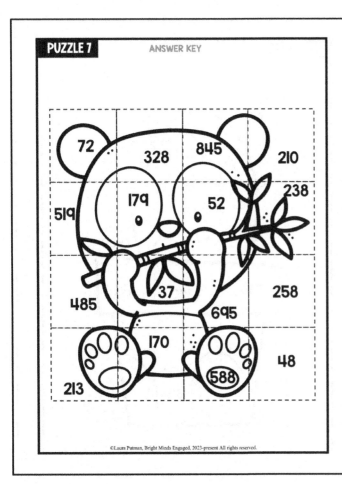
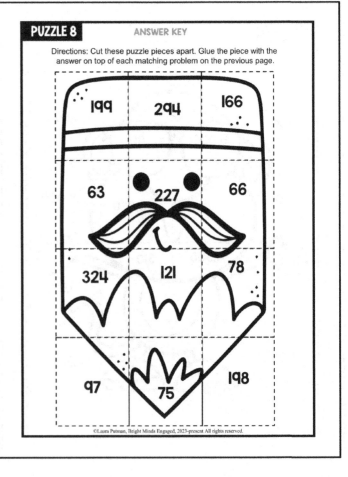

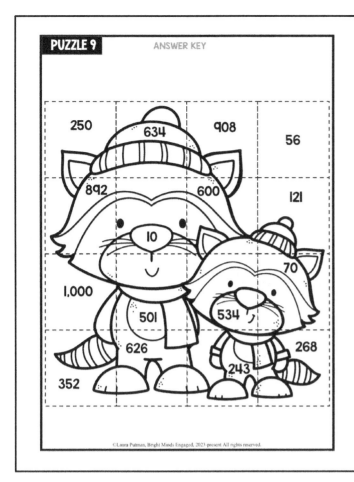
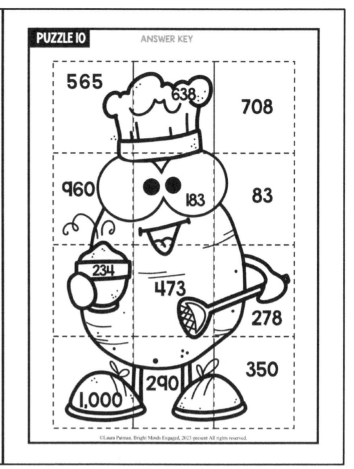
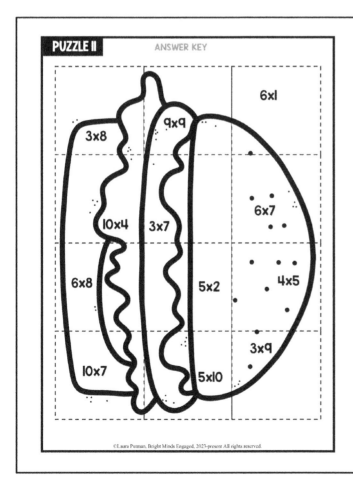
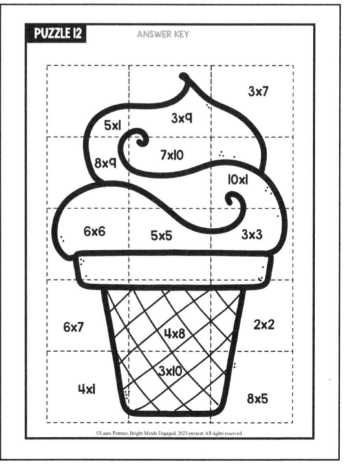

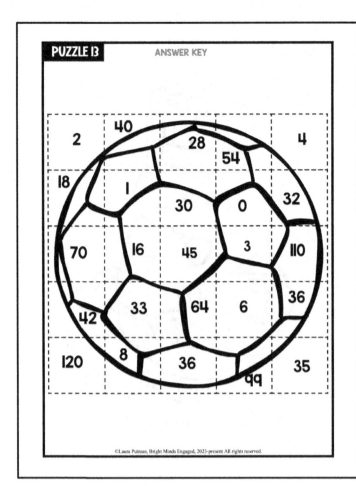
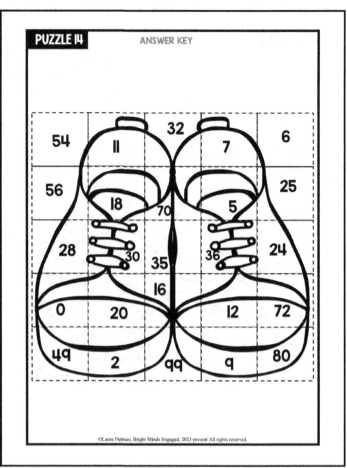
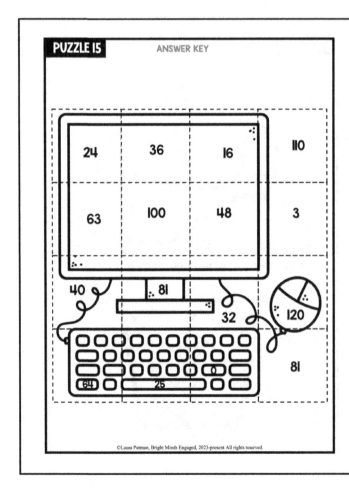
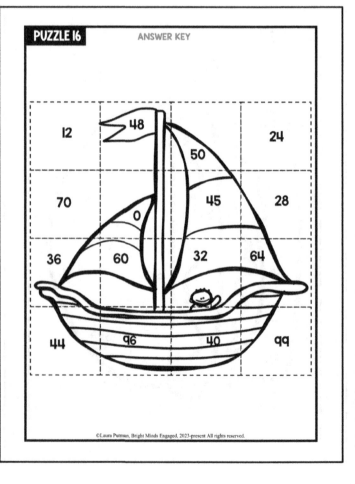

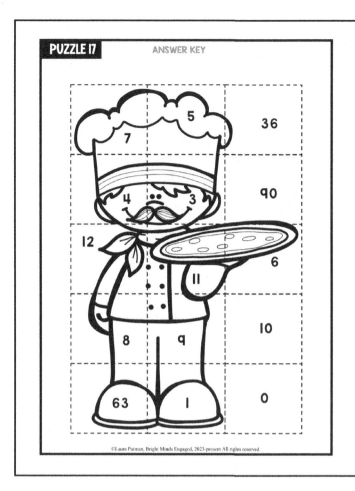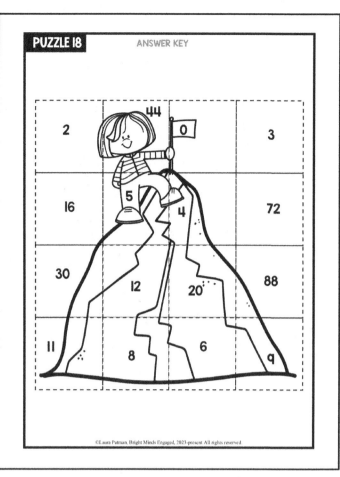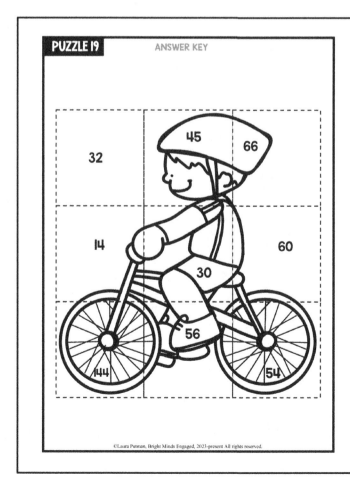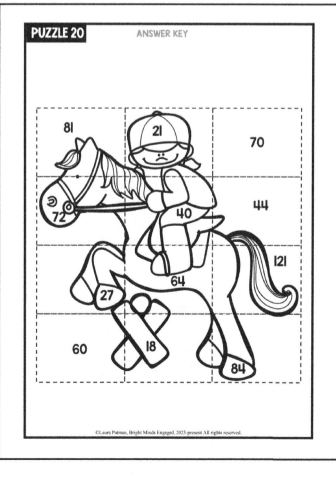

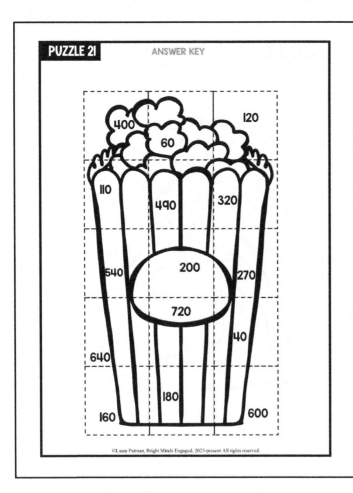
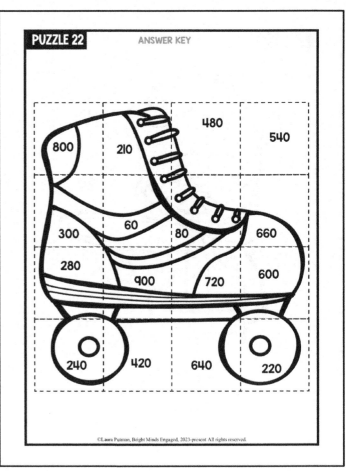
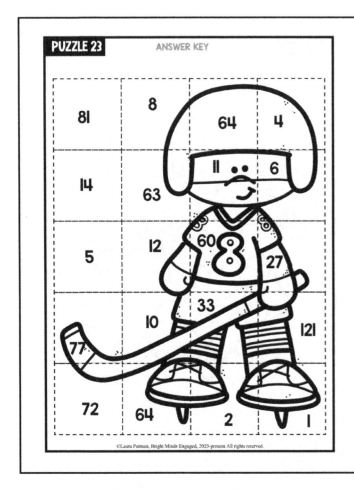
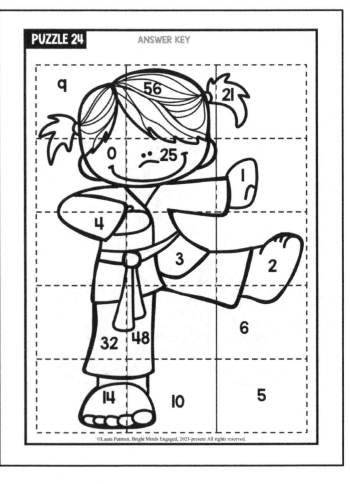

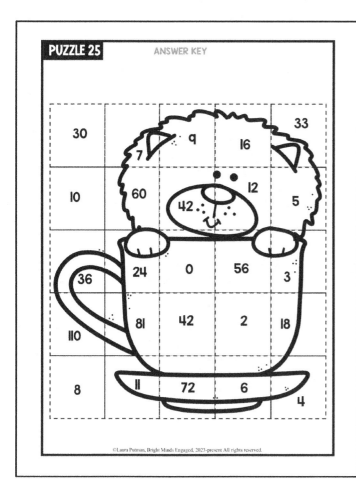
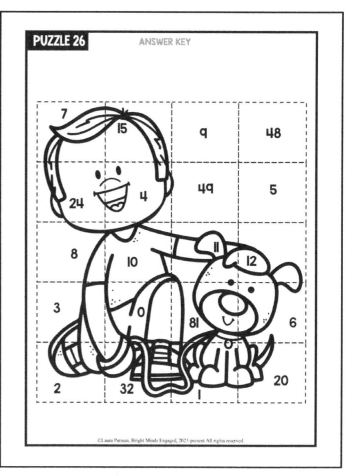
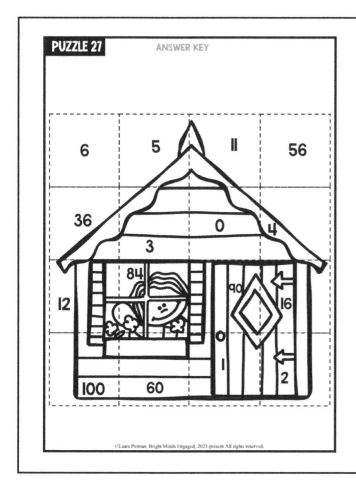

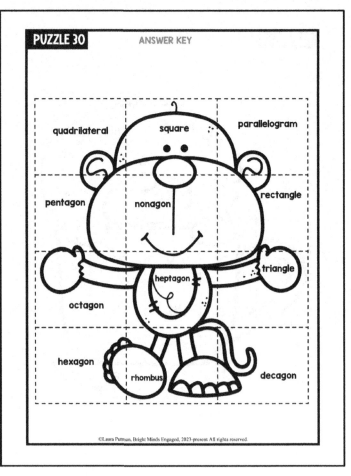
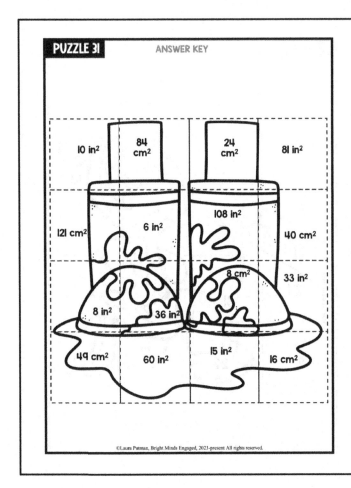
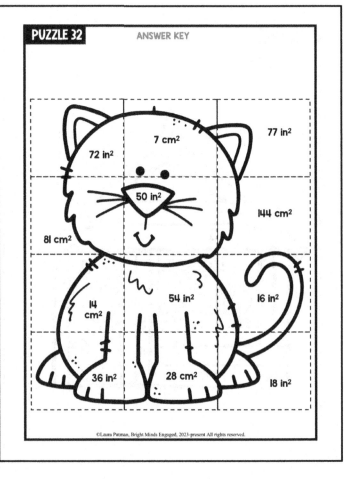

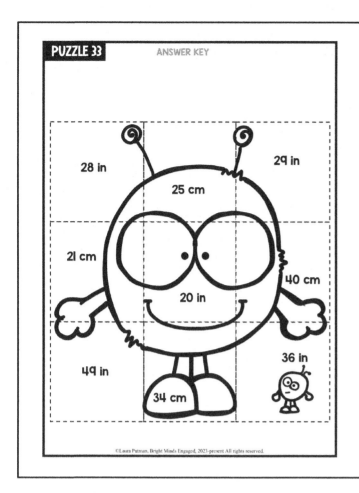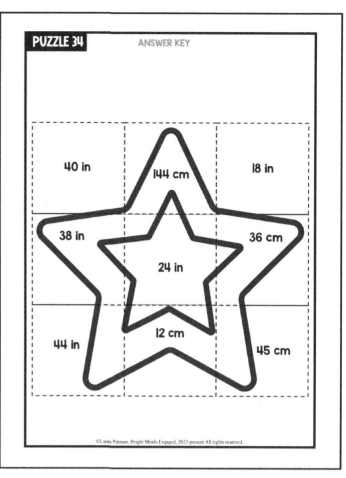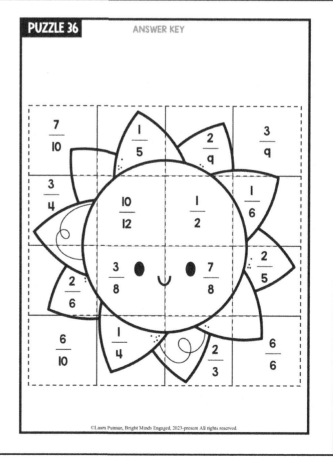

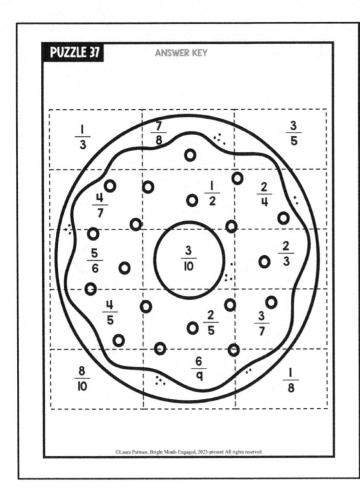
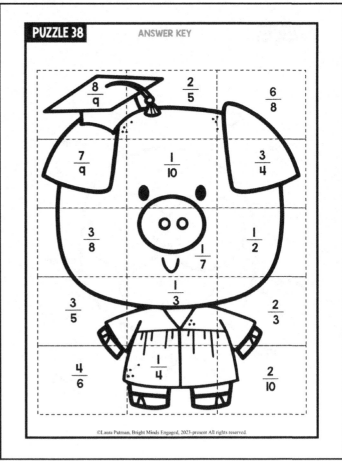
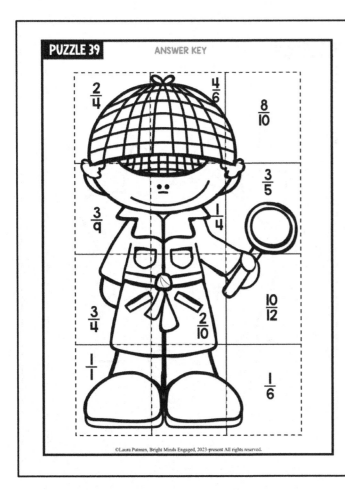
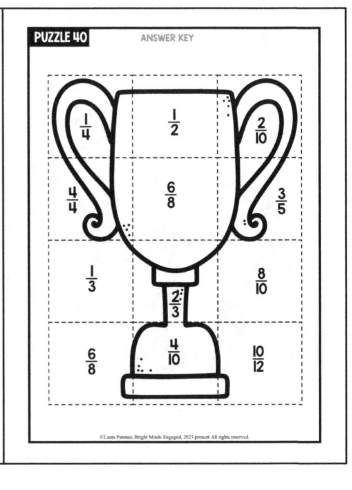

Copyrighted Materials: Laura Putman, Bright Minds Engaged, 2024-present, All rights reserved.

Scan the QR code to get a FREE sampler of my Multiplication Facts Worksheets!

Copyrighted Materials: Laura Putman, Bright Minds Engaged, 2023-present, All rights reserved.

Want a FREE multiplication facts game you can use today? Scan the QR code!

brightmindsengaged.com

Made in the USA
Columbia, SC
16 July 2025